NUTRITION FOR THE BRAIN

Feeding your Brain for
Optimum Performance

Read this book and apply what Dr. Krebs teaches and you will feel better and think more clearly than the other people in your life... Then you need to get them to read it too.

Socrates said 'Know thyself.' Reading Dr. Krebs' enjoyable and informative book will help you to better 'know yourself' and teach you much about why you feel and act the way you do. But even better, it gives you guidance about how to improve your mental function for a happier and more productive life.

Employers, teachers, coaches, any leaders, will want their group to understand and follow the common sense principles that Dr. Krebs' book teaches. The results will be increased productivity and a happier environment.

Walter H. Schmitt, DC, Diplomate in AK and Diplomate in Chiropractic Neurology
Chapel Hill, NC 27514
USA

I highly recommend this book to anyone who has ever been confused about exactly what a 'healthy' diet is, or wondered how you could develop nutritional deficiencies when eating a 'balanced' diet. In a lucid, easy to read style supported by clear and uncomplicated diagrams, Dr. Krebs presents all of the basic principles of nutrition plus a wealth of information on how nutrition can help you achieve optimum performance.

Gerhard Otto, MD, Specialist in Nutritional Medicine,
Essen,
Germany

NUTRITION FOR THE BRAIN

Feeding your Brain for Optimum Performance

Dr Charles Krebs

MICHELLE ANDERSON PUBLISHING
MELBOURNE

First published in Australia 2006
by Michelle Anderson Publishing Pty Ltd
PO Box 6032 Chapel Street North
South Yarra 3141 Melbourne Australia
Email: mapubl@bigpond.net.au
Website: www.michelleandersonpublishing.com

Copyright: Charles Krebs 2006

Cover design: Deb Snibson, Modern Art Production Group
Typeset by: Midland Typesetters, Australia
Printed and bound by: Griffin Press, South Australia

National Library cataloguing-in-publication data:

Krebs, Charles T.,
Nutrition for the brain: feeding your brain for optimum performance.

Bibliography.
Includes index.
ISBN 0 85572 375 0 (pbk).

1. Nutrition. 2. Minerals in human nutrition.
3. Vitamins in human nutrition. 4. Mental health. I. Title.

612.3

Disclaimer:
This book is not intended to be used as a substitute for professional health care and assistance. While every effort has been made to provide accurate and reliable information, the author and publisher cannot be held responsible for any consequences of applying the facts and advise presented.

Foreword

This book explains what makes up nutrition, how nutrients work in our body and how the lack of nutrition has detrimental effects. Likewise, it highlights how good nutrition can remedy many chronic disease conditions increasing your health, vitality and, most importantly, help you to achieve and maintain optimal mental performance. While it concentrates on nutrition that feeds the brain, allowing you to maintain peak performance even when under stress, much of the following discussion applies equally to physical performance.

Because many people have relatively little idea about what nutrients are or how they work in your body to promote and maintain optimum function, the author attempted to provide all of the Key Principles and Concepts of Nutrition in one place. These Key Concepts are generally scattered through the nutritional literature with only a few principles presented in any single book, yet each contributes directly to how and why nutrients create and maintain our health and vitality. To assist the reader there are Concept Boxes summarizing each Key Concept as it is presented throughout the book.

In spite of the scant attention paid to nutrition and nutritional therapy in Western Medicine, there are literally thousands of scientific papers demonstrating how many of our common illnesses, diseases and dysfunctions result primarily from various nutritional deficiencies. Unlike the 'absolute' nutritional deficiencies that abound in less developed countries due to shortages of basic foods, people in Western developed countries are often 'Overfed, but Malnourished', and commonly suffer from 'marginal' nutritional deficiencies.

Marginal nutritional deficiencies exist when there is enough of a specific nutrient to meet low to normal levels of function, but which are depleted during times of 'stress' and 'peak' demands reducing the ability of both our brains and bodies to operate anywhere near their optimum level of performance. For example, over two billion people in the world today suffer chronic zinc and iron deficiencies that affect their thinking, short-term memory, and moods, as well as their immune function, and energy levels. Why marginal nutritional deficiencies can exist in developed countries with a surfeit of food is thoroughly discussed, as are the consequences of these nutrient deficiencies.

Not surprisingly, one of the primary factors creating these nutritional deficiencies are the foods increasingly eaten in the Western diet. Surprisingly, these deficiencies result in large part from a loss of much of the nutrient content of our produce due to the changes in farming and food storage and handling practices that have occurred over only the past 30 years – combined, these factors have reduced the overall nutrient content of our produce by 20 to 80%. Not surprisingly, the other major factor is the change in the types of foods people are increasingly choosing to eat.

Fast Foods and Junk Foods (as the name so accurately

suggests) are woefully deficient in many important nutrients, while being exceptionally 'rich' in energy content. These foods thus present a double threat to your health and well-being. First of all, the body's first response to a lack of nutrients is simple – eat more food! However, when the food eaten is also deficient in nutrients, it initiates another round of – eat more food! Because these foods are also incredibly 'energy dense', what do you do with all of this extra 'energy'? Since it takes up to 40 different nutrients to 'burn' this high energy content, but relatively few nutrients to turn it into fat, eating nutrient deficient Fast and Junk Foods are a major, if not the primary factor causing the epidemic of obesity spreading across the developed countries today!

Other factors such as Life Style also may play important roles in creating nutritional deficiencies. The 'on-the-go' lifestyle so prevalent today often leads to the consumption of less nutritious foods, either because they are unbalanced in their nutrient content, or because they have been highly processed reducing their nutritional quality. In counterpoint to the 'on-the-go' lifestyle, is the 'couch-potato' lifestyle so common today which is also often associated with poor food choices not supporting optimum mental function and with a lack of exercise so vital to the maintenance of our health and well-being. Exercise oxygenates the brain providing the fuel for thinking when adequate nutrients are available, the reason getting up and moving around can help bring your mind back on-line when you are doing heavy mental work.

Your genes are a factor that is often over-looked, partly because the role they play in nutrition is little understood by the majority of the population. Yet your genes determine how well you will absorb and utilize the nutrients that you do consume. Many marginal nutritional deficiencies arise due to people having different genes that code for the same proteins, and enzymes, transporter and receptor molecules are all proteins. Thus, some people have faster, more efficient enzymes, transporter and receptor molecules than other people, and thus can absorb, assimilate and utilise the nutrients in their food far more effectively than other people who have genes coding for less efficient molecules.

What nutrients are and what they do in the body is the topic of Chapter 3. While the emphasis is upon the role nutrients play in brain function, much of this discussion applies equally well to the rest of the body. Chapter 4 uses zinc, a nutrient that is commonly deficient in many people's diet, as a case study to highlight the nature and effects of specific nutritional deficiencies.

Chapter 5 introduces a whole new concept in brain function called Brain Integration, which is a new understanding of how the brain actually works. For efficient processing, the brain uses multi-plexing and parallel processing, that is uses many of the same basic processing modules, but in unique combinations with each new combination representing a new function. Thus, for thinking to work effectively, all of the different brain areas involved in processing must be synchronized in time to maintain function. A loss of integrated, synchronized neural flows disrupts the functions dependent upon these flows and thus results in a loss of access to these brain functions.

Clearly, the basis of brain integration is biochemical-nutritional, that is having sufficient nutrients to maintain the levels of neurotransmitters needed to keep these neural flows timed and moving! Reduction in neurotransmitter levels due to lack of nutrients needed to make them can lead to a total loss of brain integration and hence dysfunction, especially during times of 'peak' demand when you are under Stress!

Chapter 6 discusses what 'stress' is, and how it can affect your mental performance by causing loss of integrated brain function. Chapter 7 then covers the role of nutrition in maintaining brain integration and how having adequate nutrients available during times of 'stress' can keep you thinking and problem-solving instead of sliding into dysfunction that creates even more 'stress'.

Chapter 8 presents a nutritional approach to maintaining integration and full function when under 'stress' by providing the nutrient matrix necessary to support optimum mental performance. From the discussion of the many nutrients needed to maintain optimum function, it is clear it would be difficult to do so taking each of the individual nutrients required. Chapter 9 presents a solution to this problem, Nutriceuticals, complex nutritional formulae designed to support specific functions, in this case the integrated brain function underlying optimum mental performance.

The Appendices provide additional information on the vitamins and minerals discussed in the text, including not only the sources of each nutrient, but its function in the body, consequences of deficiency, and both the Recommended Dietary Allowance and Therapeutic Dosage Range. Appendix 3 presents a novel solution to the problem of how to provide optimum nutritional support for maintaining brain integration under Stress in the form of the nutriceutical, ThinkingAdvantage.

An extensive Glossary is provided that defines many of the more technical terms and words or important concepts used throughout the book. It is hoped that this will be a valuable tool for clarifying your understanding of both the terms and concepts presented.

So much of the happiness and contentment in our lives today depends upon how well we can handle the 'stresses' of every day life, which seem to multiply with each passing year. We all need fully functional brains capable of providing creative solutions to problems confronting us on a daily basis. The purpose of this book is to understand why it is important to provide a proper nutrient matrix to optimize our brain function, thus keeping our problem-solving abilities 'online' even when we are under 'stress'. By following the guidelines contained in this book you will be able to provide a proper nutrient matrix to optimize your brain function and keep your problem-solving abilities 'on-line' under 'stress', indeed when you need these abilities the most!

Hopefully, you will find the book both informative and enjoyable to read.

Dr. Charles T. Krebs
Melbourne, Australia

Acknowledgements

I would first of all like to acknowledge Beate Walter and Alfred Schatz of the Verlag für Angewandte Kinesiologie (VAK) who urged me to write this book as part of their Concept Series. Without their insistent prompting and guidance, this book would not have been written. I also wish to thank the Editor of VAK, Nadine Weber, for her assistance at every stage of production and for her insightful comments on the manuscript for the Concept Book, Nährstoffe für ein Lestungsfähiges Gehirn.

I owe a huge debt of gratitude to Ms Jill Innes for her superb editing skills and the many contributions she made on the layout of the book and her suggestions of where and what type of graphic was needed to increase understanding of the concepts presented. I also thank her for playing the role of my 'everyman' with regard to the technical details presented in the book to keep the flow and maintain a lay person's level through out.

I wish to thank my friend and partner, Nigel Griffith for his encouragement and support and many other people who have at various times either in discussions or by reading parts of the manuscript, made suggestions that have been incorporated into the book making it both more highly readable and informative.

Most of all, I am indebted to my lovely and loving wife,

Stefanie Maurer, for all of her encouragement at times when I felt overwhelmed, and did not see how I could ever finish the book in the face of my many other competing projects. And most of all, for being so understanding of the time I took away from my family to complete this project.

Contents

Foreword		v
Acknowledgments		iii
Chapter 1	Nutritional Deficiencies	1
Chapter 2	Reasons for Nutritional Deficiencies	9
Chapter 3	Nutrition and How it Works	40
Chapter 4	The Zinc Connection	54
Chapter 5	Introduction to Brain Integration	83
Chapter 6	Effects of Stress on Mental Performance	100
Chapter 7	How Nutrition Can Optimize Mental Performance	114
Chapter 8	Nutrition for Optimum Mental Performance	133
Chapter 9	Nutriceuticals for Optimizing Mental Performance	154
Glossary of Terms		167
Annotated List of Suggested Reading		190
References and Chapter End Notes		196
Appendix 1:	Table of Vitamins: Sources, Doses, Functions and Deficiency Symptoms and Disorders	227

Appendix 2: Table of Minerals, Sources, Doses,   236
            Functions and Deficiency Symptoms
            and Disorders
Appendix 3: ThinkingAdvantage: One Solution to   246
            Optimizing Mental Performance

Index   264

Chapter 1

NUTRITIONAL DEFICIENCIES: WHAT ARE THEY AND WHY DO THEY EXIST?

INTRODUCTION

When most people think of Nutrition, they think about the food they eat and this is, indeed, the source of most of our nutrition, yet many people are unaware of the benefits of eating a balanced diet that can greatly improve the quality of life for themselves and their family. Likewise, many people also do not understand why nutritional supplementation may be necessary to achieve optimum physical and mental performance.

Optimum mental performance depends upon maintaining integrated brain function under stress and it is nutrition that plays a major role in the performance of the brain. Different brain functions and regions require different nutrients to fuel the transmission of nerve impulses and maintain integrated function, or Brain Integration. Failure of effective neural signaling is often caused by having insufficient levels of the right nutrients to fuel this level of neural activity, especially under stress when more nutrients are needed. Indeed, recent studies show that even at rest, a major fraction of all the glucose (sugar) used in the brain is used just to maintain neurotransmitter levels for normal function.

While most people have heard about nutritional deficiencies, that is, not having enough of a specific nutrient or nutrients in your diet, they are not always sure what this really means, or how you get this problem in the first place. First of all, it must be realized there are two entirely different types of nutritional deficiencies, *absolute nutritional deficiencies* and *marginal nutritional deficiencies*.

Absolute deficiencies are based upon depletion of nutrients because these nutrients are almost wholly absent from the diet. In contrast, marginal deficiencies are just lack of specific nutrients in large enough quantities to handle 'peak' demands for these nutrients.

With *marginal deficiencies*, at normal levels of activity and function the nutrient levels present are sufficient to support basal body and brain function. However, when the level of activity or intensity of the function reaches a certain level, the body or brain just runs out of enough of specific nutrients to maintain optimum function.

> **Key Concept: Absolute and Marginal Nutritional Deficiency**
>
> *Absolute nutritional deficiencies* are based upon the depletion of nutrients because these nutrients are almost wholly absent from the diet. The most common cause of absolute deficiencies is starvation.
>
> *Marginal nutritional deficiencies* in contrast, result from the diet just not providing enough of specific nutrients to handle 'peak' demands for these nutrients, especially when under stress. At normal levels of

activity and function the nutrient levels present with *marginal deficiencies* are sufficient to support basal body and brain function. However, when the level of activity or intensity of the function reaches a certain level, the body or brain just runs out of enough of these nutrients to maintain optimum function. There are just not enough neurotransmitters produced to support integrated brain function.

ABSOLUTE NUTRITIONAL DEFICIENCY: TOO LITTLE FOR TOO LONG

Most people suffering from absolute nutritional deficiencies 'know' they have these deficiencies because absolute deficiencies are usually associated with specific nutritional diseases, such as scurvy resulting from an absolute deficiency of vitamin C. This is the reason the British sailors were called 'limeys', as even before vitamin C was discovered, the British Admiralty had observed that having the sailors eat limes on long voyages prevented scurvy. Hence all British sailors were required to eat a lime or the juice of a lime every day on a long voyage but, of course, they needed a bit of grog or rum to wash it down!

Likewise, the discovery of vitamins resulted from observing chickens fed on rice that had been milled to remove the husk and nutrient-rich outer layer. A researcher and medical doctor, Chritiaan Eijkman, was living in Indonesia during WW II because of food shortages could at times only get milled white rice to feed his chickens, whereas before the war he had feed them whole brown rice. To his surprise, even though the chickens appeared to be getting 'enough' to eat, they began to show abnormal behaviour and became sick, often becoming so weak they were unable to walk. He observed the same symptoms among the natives who before the war were healthy. So he reasoned that since the only difference in both the chickens and the natives diets was the husk or outer layer of rice that had been milled away during processing, this must contain something chickens and people needed to be healthy.

When he then supplemented the diet of the chickens with the outer layer of the husk milled away in processing, they once again became healthy. He called this new type of dietary requirement for healthy function a 'vitamin'. Later when the husk was analysed, the active compound was identified as the molecule Thiamine. Thus today Thiamine is known as Vitamin B_1, and deficiency of vitamin B_1 is known to lead to muscle weakness.

The origin of absolute deficiencies is generally obvious, the absence of or far too little of a particular nutrient in your diet, including macro-nutrients. This usually results from just not eating enough food, called starvation, or from eating only foods that are almost totally devoid of these particular nutrients.

The bountiful amounts of most basic foods in Western societies means that absolute nutritional deficiencies are relatively rare in the developed countries. In contrast, the lack of sufficient food in many developing countries often leads to absolute deficiencies.

Marginal Nutritional Deficiency: A Hidden Problem

While absolute nutritional deficiencies announce their presence by creating observable dysfunction, marginal nutritional deficiencies often go unnoticed. Since marginal deficiencies permit normal or low levels of activity and function, and only result in decreased mental and physical performance when we are under stress, they are often not acknowledged. Rather, the dysfunction caused by these marginal deficiencies is often attributed to factors other than nutrition such as being stressed or being physically tired, which are actually the main symptoms of marginal nutritional deficiencies.

In marked contrast to people in developing countries suffering absolute nutritional deficiencies due to lack of food, people in developed countries seldom suffer from macro-nutrient deficiencies, but commonly have marginal nutritional deficiencies for several reasons including life style and food choices. Many people eating a typical Western diet are overfed, but malnourished leading to *marginal nutrient deficiencies.*

This results partly because much of the diet is made up of highly processed foods rich in simple sugars that provide adequate energy, but virtually 'no' nutrients. While pure white refined sugar is indeed one of the purest things you can eat being 99+% pure sucrose, it is almost totally lacking in nutrients. During the refining process 64 nutrients are removed or destroyed: all potassium, magnesium, calcium, iron and manganese are removed and the vitamins A, D, and all of the B vitamins are destroyed. All amino acids, vital enzymes, unsaturated fats and all fiber are eliminated.

Sugar is not a 'bad' food, in fact it is an essential food, as the body uses glucose to produce energy and the brain runs entirely on glucose. However, many people may consume up to 40% or more of their total kilojoules (calories) from sugar, yet often think they don't eat much sugar – 'I only use a teaspoon of sugar in my coffee/tea!' However, there is considerable sugar 'hidden' away in today's Western diet, especially in many processed foods and soft drinks. What many people are unaware of is that soft drinks contain as much as 10 teaspoons of refined sugar per 12 ounces (375 ml) can or bottle, or 10 grams per 100 grams, and many commercial fruit juices with added sugar have similar amounts of total sugar.

Sugar, in addition to high carbohydrate breakfast, has been shown to increase deviant behaviour in hyperactive children, while the same amount of sugar consumed with a high protein breakfast, had no significant effect. In another study, a moderate amount of sugar (28 grams for a 20-kilogram child), when eaten with a meal balanced in fat, carbohydrate and protein, actually improved classroom performance by decreasing reaction time, errors and activity levels.

In 1915 the average American consumed between 15 and 20 pounds (7 to 9 kilos) of sugar per year, and this was, of course, mostly unrefined sugar (what we call 'raw' sugar today) that contained at least some nutrients. Today, however, the average American consumes more (and often much more) than his or her body weight per year in refined sugar, plus another 20 pounds (almost 10 kilos) of corn syrup. Since refined sugar only provides pure energy, the body must 'borrow' vital nutrients like calcium, magnesium and potassium from healthy cells to metabolise it, and then

NUTRITIONAL DEFICIENCIES

convert any excess sugar into fat. Consequently, excess refined sugars deplete the body's stores of valuable vitamins and minerals. In *Lick The Sugar Habit*, Nancy Appleton sights 111 scientific references supporting 76 different negative effects excessive intake of white refined sugar can have on the body.

In addition, food processing also overtly destroys B-vitamins including vitamin B_6 and folates, which affect the brain and central nervous system function. For example, a Junk Food diet is deficient in thiamine (vitamin B_1), which has been associated with increased aggression. According to Dr. Mercola, (**www.mercola.com**) about 90 per cent of the money Americans spend on food today is spent on processed foods.

The Western diet has also been shown to be deficient in iron and zinc, resulting in brain dysfunction and learning impairment, including reduced IQ. Two recent scientific studies have shown that individually, zinc and iron deficiencies are associated with Attention Deficit Hyperactivity Disorder (ADHD), and that supplementation with either zinc or iron resulted in improvements in the ADHD symptoms. This is because iron and zinc are co-factors for noradrenalin, dopamine and serotonin synthesis in the brain, and hence control production of the vital neurotransmitters that affect mood and behaviour, and zinc is also an important co-factor in many enzyme systems supporting both body and brain function, especially thinking and memory.

Junk and Fast foods are commonly lacking in adequate amounts of many nutrients, as convincingly demonstrated by a young healthy film-maker, Morgan Spurlock, who made a documentary using himself as the central character, *Super*

Size Me. He first had a complete set of physical examinations including blood tests and was found to be very fit and healthy. As an experiment, he then proceeded to eat all three meals every day at McDonalds. In three weeks he had gained 15 kilograms, was lethargic, and moody, and new blood tests showed major liver problems, especially with detoxification. Interestingly, many of the symptoms he developed are all classic symptoms of marginal nutritional deficiencies?

You might ask 'How could someone eating so much food end up in such strife after only three weeks of consuming this food?' Yet the truth is he did end up in such strife – one of the examining doctor's stated he did not think it was possible to develop such a toxic liver in so short a time. How can this be? How can someone be eating what is so common in many Western diets end up with marginal nutritional deficiencies?

If you are eating truly 'healthy' food, you should be able to eat it every day for the rest of a long healthy life!

The answer is that too much of this food contains large portions of sugar, fats, oil and carbohydrates (macro-nutrients), but not enough nutrients, especially the micro-nutrients such as vitamins and minerals, and fibre to maintain good health.

Indeed, one of the most commonly noticed results of broad based vitamin and mineral supplementation is an immediate increase in energy levels, clearer thinking and often better memory. And this is generally with people who think they already eat a reasonably 'good' diet!

How can this be? How can someone be eating a reasonably 'healthy diet' (according to most Western standards) yet end up with marginal nutritional deficiencies that decrease physical and mental performance?

Chapter 2

REASONS FOR NUTRITIONAL DEFICIENCIES

INTRODUCTION

In the developed world people live amidst an abundance of foods of all types that are basically available all year-round. Just take a stroll through any Supermarket in any Western country and you will see this. Yet many people in these countries suffer from nutritional deficiencies. But why should so many people in Western developed countries with a surfeit of food still have nutritional deficiencies?

There are three main factors that cause marginal nutritional deficiencies, even with adequate food available:

1. **Diet:** Lack of sufficient nutrients in the food you eat and the real meaning of RDAs and DVs;
2. **Lifestyle:** Choice of foods you eat; and
3. **Genetics:** The genes you inherited determine how effectively you are able to absorb, assimilate and utilize the nutrients you do have in your diet.

DIET: LACK OF SUFFICIENT NUTRIENTS

While most people automatically assume the food they eat is nutritious, this is no longer true of a good deal of the food we consume. This results from a variety of factors including the fact that a great deal of the minerals have been stripped from our soils by intensive, chemically based farming, poor land management practices, genetic modification and agricultural practices that 'speed up' production and extend the 'shelf-life' of our food crops.

FROM FARM TO MOUTH – WHAT'S MISSING?

In 1936 the US Department of Agriculture released a report recommending that Americans should consider supplementing with minerals because the US soils had become so depleted by the farming practices over the preceding 200 years. Do you think the soils have become less deficient in minerals after another 70 years of even more intensive high fertiliser farming – a practice known to further strip nutrients out of the soil? The soils of Europe have been intensively cultivated for millennia and are now in a similar shape, partly because age-old sound sustainable farming practices like fallow fields have been abandoned over the past 50 years in favour of high intensity chemical farming.

Also, current factory farming practices may greatly reduce the nutrients available in our food. Many crops like fruits and vegetables are routinely harvested early, so they ship better and have a longer shelf life. Then they are often sprayed with ethylene gas, or other ripening agents, to force-ripen them off the vine or tree. You can often see signs of this spraying in the oranges at your local supermarket with bright orange on one side and green on the other.

While this may give an appearance of nutritious looking fruits and vegetables, it is often only skin deep, because the bulk of the nutrients are deposited in the fruits and vegetables during the final stages of ripening on the plant or tree. The whole reason for a plant to make a fruit or vegetable is to have its ripened seeds widely distributed, and to this end much of the nutrients are withheld in the roots until the final week or even days before they are truly ripe and the seeds ready to be dispersed. Thus, picking fruit and vegetables a week or more before they are fully ripe deprives the consumer of much of their natural goodness.

While making these fruit and vegetables ship and store far better than fully ripe produce, a big boon for the supermarkets, they may leave you nutritionally deficient. Likewise, the practise of Modified Atmosphere Packaging where produce like lettuce and salad greens are packaged under a nitrogen or carbon dioxide atmosphere may indeed inhibit rotting and decay because of the absence of oxygen for up to several weeks, it often leaves the produce deficient in antioxidants and nutritional quality. Of course, then there is the common practice of holding produce, especially fruits and vegetables, in cold storage at 4 degrees Celsius under a CO_2 atmosphere, often for months. While this may make produce available at unseasonable times of the year, this long-tem storage has done little for the nutritional value of the food by the time it gets to your mouth.

Recent research in Britain by John Thomas showed that between 1940 and 1991 levels of essential minerals such as iron, zinc, magnesium, potassium and copper important for the body's biochemical balance have declined by between 25% and 75% in a number of fruit and vegetables. It also showed that the proportion of sugar has doubled in fruit such as apples and pears over the same period – partly to satisfy modern tastes. For instance, American government research found that apples can now comprise up to 15% sugar compared with 8%–10% three decades ago. Similar increases have been reported in a variety of other species including pineapples, pears and bananas.

These findings are supported by a study reported in the British Food Journal by Anne-Marie Mayer, a nutrition researcher at Cornell University, who found similar changes in the nutritional content of 20 fruits and 20 vegetables grown in Britain between the late 1930s and the 1990s. 'There were significant reductions in the levels of calcium, magnesium, copper and sodium in vegetables and in magnesium, iron, copper and potassium in fruits. The greatest change was the reduction in copper in vegetables to less than one-fifth of the 1930s level,' Mayer said. Both researchers link the decline to the intensification of farming and that modern agricultural chemicals and techniques could be depriving plants of minerals.

It may be purely a coincidence that the National Diet and Nutrition Survey of Britain recently found that the blood plasma of a 25% of British men and 33% of women was iron-deficient and that many people may also be deficient in nutrients such as selenium and vitamins C and B_{12}, but it is an interesting statistic in light of recent knowledge of general decline in the nutritional content of food.

A recent German report sadly showed an even more disturbing trend. Between 1985 and 2002, there were decreases in the nutrient levels in most fruits and vegetables tested. For instance the calcium levels in many of the common fruits and vegetables dropped by more three-quarters, Vitamin B_6 levels by 80 to 90%, and Vitamin C levels by up to 80% less – a trend amazingly similar to that observed in England.

Similar results were obtained by a German company that collected fruits and vegetables from markets in Hamburg and Horneburg in May 2004. They found that the overall mineral content decreased by 33% with some important minerals like zinc by 72% when compared to 1994 levels. Vitamin B_1 varied from 78% more than 1994 values in Celery to 89% less in Fenchel, and Vitamin C was decreased in all fruits and vegetables tested with an average decrease in of Vitamin C content of 73%.

What this means practically is that for you to consume the current RDA of vitamin C in 1985, you would only needed to eat 125grams of strawberries, but by 1996 it would have been 500grams, and by 2002, almost a kilogram. Likewise if you were to get your RDA of Vitamin C eating bananas, you would only need to one banana in 1985, but 8 bananas by 1996, and a dozen bananas by 2002.

However, researchers emphasized that fruit and vegetables were still far better than processed foods in terms of nutrient quality, but warned that continued falls in nutrient levels, and rises in sugar would be a problem.

> **Key Concept: Foods today are often Deficient in Important Vitamins & Minerals**
>
> *Nutrient Deficiencies in our Food:* Studies have shown that much of the food available, particularly in Supermarkets where it is grown on 'factory farms' is lacking in many vital Vitamins and Minerals. Several studies have shown that in the past 20 to 40 years the nutrient levels of many common foods consumed today

> decreased between 25% and 80% in important vitamin and mineral content. This makes it difficult today to get all of the nutrients you need from your diet alone, unless you consume biodynamic or organic grown foods that *do* have much higher nutrient levels.

WHAT FOOD SHOULD AND SHOULDN'T CONTAIN: NUTRIENTS AND NOT TOXINS!

In a study of vegetables and fruits produced by biodynamic, organic and conventional factory farming in Australia in the 1980s, there were large differences in the nutritional content of the foods. For instance, tomatoes grown biodynamically and organically had 1000 mg of vitamin C per 100 grams of flesh, while factory farmed tomatoes had only 100 mg of vitamin C, 10 times less of this vital nutrient! There were similar differences in the mineral content of the other fruits and vegetables analyzed. Would you eat 10 of these factory farm tomatoes to make up the difference? In the 2004 analysis of German tomatoes there was even less Vitamin C with only 20 to 29 mg per 100 grams, suggesting things are getting worse, not better.

In a recent Danish study, rats were fed diets containing either totally organic grown food, low fertiliser grown food, or conventionally grown food, and then their health monitored throughout their entire lives. The results were remarkable as researchers found rats that ate organic or minimally fertilized diets had:
- Improved immune system status compared to rats that ate conventional diets.
- Better sleeping habits.

- Less weight and were slimmer than rats that fed on other diets.
- Higher vitamin E content in their blood, an important antioxidant (for organically fed rats).

Even though the experiment clearly demonstrates the positive effects of organically grown foods compared to conventionally grown foods on the health of rats, the results cannot be directly correlated to humans. However in many other studies such as the toxicity of drugs, the results from rat trials are considered significant and these results certainly suggest a similar trial be repeated with humans – but perhaps not for their entire lifetime.

In conventional farming there are now over 400 chemicals routinely used to kill weeds, insects, and other pests that attack crops. For example, Cox's apples in the UK can be sprayed up to 16 times with 36 different pesticides. Current farming practices using all of these herbicides and pesticides leave food contaminated with these toxins that are not easily 'washed' away. Sir John Krebs, Chairman of the UK Foods Standards Agency, recently stated that 'Organic foods contains fewer residues of the pesticides used in conventional agriculture, so consumers who wish to minimise their dietary pesticide exposure can do so with confidence by buying organically grown food.' A 2002 study of preschoolers in Seattle showed that children who ate a conventional diet had nine times the levels of pesticides in their urine as their counterparts who ate organically grown foods, and another study in 2003 replicated these results.

A benchmark study of the pesticide and Industrial toxin contaminants in the umbilical cord blood of 10 healthy newborn infants is both surprising and distressing. The study

focused on the cord blood, as this mirrors the mixture of chemicals the baby has been exposed during development. The following toxic chemicals were found in infant's cord blood:
- 76 chemicals that cause cancer
- 94 chemicals that are toxic to the brain and nervous system
- 79 chemicals that cause birth defects or abnormal development in animals

The report, Body Burden – The Pollution in Newborns, by the Washington, D.C. based Environmental Working Group, detected 287 chemicals in the umbilical cord blood of newborns, including seven pesticides – some banned over 30 years ago in the United States. Scientists blame the presence of these banned pesticides in the babies' blood on the fact that many of these compounds take decades to break down and some are still used in foreign countries, which export produce to the United States.

Although the amounts of some chemicals detected were extremely low, the results are still troubling to experts, since no one knows how much of a given chemical – much less a mixture of these chemicals – could affect a human fetus. What research exists has shown that chemical exposure in the womb can be dramatically more harmful than later in life. In 2003, the US EPA updated its cancer risk guidelines, finding that carcinogens are 10 times as potent to babies than adults and that some chemicals are up to 65 times more potent in children. While the EPA sets maximum safe exposure limits for these toxins, the research behind these tolerances came from studies of 'healthy men' in the middle of their life – not pregnant women, newborns or young children!

Aside from this pesticide contamination, conventional

produce tends to have fewer nutrients than organic produce on average, with organic produce having generally 20 to 30% more nutrients than conventional produce that may have up to 80 per cent less of some nutrients than found in organic produce. Independent studies of organic crops have found significantly higher levels of 21 nutrients including 27% more vitamin C, 21% more iron, 29% more magnesium and 14% more phosphorus, and significantly less nitrates (a toxin) than conventional produce.

In addition, a Danish study showed organic crops to contain from 10 to 50% more antioxidants than conventionally grown crops, and higher levels of antioxidants appear to promote better health and have an anti-aging effect. Compare this to a reduction in antioxidants found in conventionally grown lettuce stored under Modified Atmosphere Packaging. When people consumed this lettuce, it was shown to actually lower the blood levels of antioxidants. And of course it's not only how the food is grown, but also how it is cooked. A recent study showed that while traditional steaming left the antioxidant flavonoids almost untouched, microwaving virtually eliminated all of them.

Researchers have long suspected that dramatic changes in agriculture over the past 60 years could be changing the quality of the produce. However, the short-term benefits for farmers such as greater productivity, consistent quality and a wider range of varieties, and produce available at unseasonable times of the year, meant that these concerns attracted little attention.

Likewise, genetic modification of the original foods, either by biotechnology or selective breeding for specific traits may also play a role in marginal nutritional deficiencies. The high yield rice touted as the Green Revolution was actually a disaster for millions of people – 'Why?' Because while it provided additional essential carbohydrate and filled many more bellies, it was almost totally lacking in iron, yet it replaced the original lower yielding crops that were rich in iron.

So, over time, whole populations became iron deficient. The people eating this Green Revolution rice, while better nourished with macro-nutrients, became progressively deficient in iron, resulting in lowered energy levels, lower IQs and lower intellectual function. This is now being addressed by inserting a gene to increase the iron content of the next generation of Green Revolution rice, but whether this will be enough to offset the loss of traditional iron sources remains unknown! Likewise, GM 'golden rice' that has been genetically modified for higher Vitamin A levels actually provides only provitamin A, and it is not known whether this will be destroyed by cooking or absorbed by the body.

A recent Japanese study suggested a link between increased violence in animals and potentially children to an herbicide routinely required to grow Genetically Modified crops. Glufosinate ammonium and glyphosate are two herbicides used with the herbicide-tolerant transgenic crops that currently account for 75 per cent of all transgenic crops worldwide. Part of the genetic modification is insertion of a gene to rapidly metabolise these specific herbicides permitting high concentrations to be sprayed on the crops to eliminate all other competing plants. It appears that metabolites of Glofosinate are converted into hormone-like substances in the body that may increase aggressiveness in animals, and high levels of this metabolite have been found in people.

A report in May of 2005 by the Independent, a British-based news source, revealed secret research done by GM food giant Monsanto that compared the biological effects of eating GM corn versus naturally grown corn on rats. According to the 1,139-page report:
- Rats fed GM corn had smaller kidneys and variations in the composition of their blood – raising concerns that human health could also be adversely affected by eating such foods.
- Health problems were unseen in the rodents fed non-GM food.

Based on the findings, doctors speculate that the changes in the blood of the rodents could imply that the rats' immune systems had been damaged, or that a disorder such as a tumor had grown and their systems were trying to fight it. So it is unclear whether GM food provides the same nutritional quality as non-GM foods, and what unforeseen side-effects of consuming GM foods might be.

Recommended Dietary Allowance (RDA) or Daily Value (DV)

Another reason why someone who might by eating a reasonably 'healthy' diet may still end up with marginal nutritional deficiencies is because most Western standards are based upon the concept of Recommended Dietary Allowance (RDA) or Daily Values (DV) of a nutrient. This value is said to represent the amount of a specific nutrient needed every day to be 'healthy' as opposed to being 'unhealthy'. Western doctors and nutritionist often make the blanket statement – 'All you need to be healthy is to eat a balanced diet!', where a 'balanced diet' means the ingestion of 100% of the RDAs of all of the major nutrients.

The Meaning of the Recommended Dietary Allowance or Daily Value and Marginal Nutritional Deficiency

The average person is familiar with the concept of the Recommended Dietary Allowance (RDA) or Daily Values (DV) of a nutrient, usually expressed as a percent of RDA on cereal boxes and vitamin and mineral bottles. But exactly what is an RDA or DV and what does this mean?

To understand this, you first have to understand the origin of the RDAs used today. RDAs were originally the *absolute minimum amount of a specific nutrient – vitamin or mineral – that you needed in order not to develop an overt nutritional deficiency disease.* The original RDAs were established following WW II when millions of people in Europe were starving, and the United States had surplus food though not enough for everyone. So research was done to find out *how little of each major nutrient you could consume just to stay alive* and not die or become overtly ill from a nutritional deficiency disease.

So the initial *RDAs represented a disease deficiency guideline* – how much of each nutrient you had to ingest not to develop a deficiency disease – and therefore, had little to do with the levels of nutrients were required for optimum health and function. Over the years the amounts of nutrients needed to meet the RDA has increased upon the basis of further research, *but is still dominated by the basic concept that absence of overt nutritional disease represents health!*

Witness this quote from a recent chemistry textbook: 'The National Research Council of the National Academy of Sciences (in the US) has developed recommended dietary allowances (abbreviated RDA) for vitamins (and minerals), set at levels *that provide adequate nutrition for healthy individuals.*' Nothing could be further from the truth! (Parentheses and italics added by this author.)

Even though this concept that absence of overt disease or sickness represents health has been known to be untrue scientifically for over 50 years physiologically, it is still stated by most doctors. In this view, there is a line representing Sickness on one end and Health on the other end with a bar in the middle – when you cross the line from the Health side to the Sick side you become Sick, otherwise you are Well (See Fig. 1a next page)! This view is totally untrue, because if a disturbance in the physiology persists over time, the body compensates for this imbalance, which while more energetically expensive than true homeostasis representing Wellness, prevents you from being overtly Sick. So a true representation of Health or Wellness to Sickness scale is a long line with a bar almost at one end above which you are truly Well, and a long space to a bar near the other end, below which you are overtly Sick or have an identifiable Disease (See Fig. 1b next page).

All consuming the RDAs of the major nutrients do is keep you above the Sick bar in a state of permanent inefficient compensation, they do not make you Well. However, in this state of inefficient compensation, you often develop 'symptoms' that can conveniently be treated with drugs. The amount of nutrients you need to remain above the Well bar at the other end of the Health scale based on many studies are

Sickness – Disease	Healthy – Wellness
Less than RDAs	RDAs

Figure 1a Old Model of Health and Wellness: where the absence of Sickness or Disease is considered to be Well, and RDAs are believed to keep you healthy.

Less than RDAs	RDAs	Many times the RDAs
Sickness	**Not Sick Yet**	**Healthy**
Disease	**Physiological Compensation**	**Wellness**

Figure 1b A True Model of Health and Wellness: There is a large area of physiological compensation between Wellness and Sickness, a whole range of Not Sick Yet. RDAs were only developed to prevent absence of overt Sickness or Disease, not to make you Healthy or Well. Thus while the RDAs may keep you over the Sickness line, they are unlikely to keep you truly Well and Healthy.

generally 5 to as much as 50 or more times the RDAs. Clearly the RDAs have more to do with Absence of Disease, and than with Health or Wellness.

Thus, initially the RDA for vitamin C was set at 15 mg per day, the amount you need to prevent the overt deficiency disease scurvy, so your gums wouldn't bleed and your teeth wouldn't fall out. The RDA was then raised to 30 mg per day and more recently raised again to 75 mg per day for women and 90 mg per day for men. However, research conducted over fifteen years ago by Dr. Ames and his colleagues showed that even the most recent RDA for vitamin C of 75 mg per day (almost 5 times the original RDA) is still about 5 times less than the amount of vitamin C required to maintain healthy sperm in men.

The highest dose he tested was 250 mg per day, almost 10 times the RDA at the time, and thus considered a 'megadose' from the Western medical perspective of the RDA. However, his data showed that even this 'megadose' did not provide maximum protection for sperm, and the actual optimum value was somewhere well above this 'megadose'. This was a 'megadose' only from the perspective of the then accepted RDA. However, from the perspective of sperm 'health', he had not yet found the *actual* RDA for vitamin C!

Again and again, scientific studies have shown the RDAs for a variety of nutrients to be between 5 and 50 times below the concentration of these nutrients required for maximum protection from chronic disease, like heart disease and colon-rectal cancers. That is, RDAs and DVs are between 5 and 50 times below the concentration of nutrients needed to maintain our long-term health and prevent marginal nutritional deficiencies. *So clearly the amounts of nutrients required for optimal body and brain function far exceeds the absence of disease guidelines of the RDAs and DVs.*

This is why you can suffer from marginal nutritional deficiencies even though you are taking in or ingesting the RDA of the deficient nutrients. By consuming *and* absorbing the RDA of a nutrient, which is not the same thing, it will indeed prevent you from developing a nutritional deficiency disease, but may still result in marginal nutritional deficiencies with regard to optimum health.

For instance, if you take the current RDA of 75 to 90 mg of vitamin C a day you will indeed prevent scurvy, the vitamin C deficiency disease. But this does not mean that you

> **Key Concept: RDAs are an Absence of Disease Guideline, *not* a Health Guideline**
>
> RDAs were developed to prevent overt disease from nutritional deficiency, and hence were the minimum amount needed to prevent disease. The amount of a nutrient needed for optimum health is generally 5 to 10s of times or more than the RDA, which is usually still 10s to 100s of times or more below toxic levels.

have adequate levels of vitamin C for optimum health, that is, enough to strongly support immune function, keep your sperm healthy or to effectively scavenge free radicals produced when you are under high levels of stress.

Thus even consuming what is today considered a 'healthy diet' with 100% of the RDA for each nutrient, you may well develop marginal deficiencies especially in stress situations in which more nutrients are needed to handle 'peak' demands. At normal levels of activity and function your nutrient levels may be sufficient to support basal body and brain function, and all appears well.

However, when 'stress' comes on-line and the level or intensity of brain and/or body activity reaches a critical level, the functions in the body or brain dependent upon these nutrients simply run out of enough nutrients to maintain optimum function. To use a factory analogy, even though a big order has just come in, and the production line should be running at full speed, the products cannot be made any faster due to the lack of one or more specific materials needed to make a few key parts.

In addition, RDAs do not identify which nutrients should

be taken together as each is determined separately. Because the whole Western medical thinking is based around drugs, that is, on individual 'active' ingredients, the great majority of nutritional research has been to investigate how specific individual nutrients affect body or brain function – e.g. 'take Vitamin B_6 to help your memory!' This assumes that Vitamin B_6 acts by itself and doesn't require a 'matrix' of other nutrients for its effect. It also ignores the fact that if one of the ancillary nutrients in the 'matrix' is also deficient, taking Vitamin B_6 may produce little of no effect.

How individual nutrients interact is seldom overtly investigated and the concept of synergistic action, where one nutrient enhances the effect of another, have rarely been studied. Yet nutrients in food are almost always present in various synergistic combinations. For instance, pure ascorbic acid (vitamin C) is poorly absorbed with only about one third of what was ingested absorbed into the blood, and the rest going down the toilet. But ascorbic acid in the presence of bioflavonoids (plant derived organic molecules) is rapidly absorbed across the gut, and interestingly, all fruits that contain high levels of ascorbic acid also contain bioflavonoids.

Equally as important, RDAs do not consider the *form* the nutrient is taken in, but only the nutrient itself – e.g. Zinc is Zinc and one form is as good as another. However, the different forms of the same nutrient vary significantly not only in the dynamics of their uptake and utilisation, but also where in the body they are used. For instance zinc has at least 6 different transporter molecules, some only found in one tissue. The brain uses a specific zinc transport molecule found no place else in the body, and this molecule transports zinc gluconate in preference to all other forms. RDA's for zinc do not consider the types of zinc required by different parts of the body, what quantities of the various types of zinc are required, or what other nutrients must be present in order for zinc to be effectively absorbed and utilised.

And while zinc sulfate and zinc picolinate are cheap and easy to make and widely used as supplements, they may be poorly utilised in the body. While supplementation with zinc sulfate may indeed raise the levels of zinc in the blood, it is poorly transported into cells and thus has little effect of intracellular levels unless high concentrations are used, and high levels of zinc sulfate are associated with stomach upset. Likewise, zinc picolinate while believed to be well absorbed, picolinic acid is not produced in nature in appreciable amounts, and hence there may not be a zinc transporter molecule for this compound to get zinc into the cells.

Lifestyle: Choice of foods you eat

With today's fast paced lifestyles, many people, including many professionals, just do not have the time, or should we say – do not take the time, to eat properly, even though they know they should do so. The ubiquitous presence of convenience and fast foods that generally lack adequate concentrations of many key nutrients certainly contribute to marginal nutritional deficiencies experienced by these people.

The typical Western diet, experts suggest, is too high in sugar, refined foods, salts and fats and too low in fibre. A recent study showed that a single meal of fast food could provide up to 50 grams or 10 teaspoons of fat, accounting for 50 per cent of the total calories in the meal. The same meal may also supply more than half the amount of salt, but less than 15 per cent of the fibre recommended per day.

Other studies have shown that even people consuming a normal Western diet do not consume the recommended amounts of fibre, probably due to the highly processed nature of most Western foods. Eighty-five per cent of Americans surveyed ate less than the recommended servings of fruit and vegetables every day, and 3% ate no produce. One student who ate a primarily junk food diet of cheese, crackers, soda, cookies, chocolate and water, along with no fruit and vegetables whatsoever, ended up with scurvy, a condition brought on by vitamin C deficiency. Because of the ready availability of these processed foods, many people make the 'choice' to eat these processed foods, rather than prepare and eat more wholesome foods containing normal amounts of fibre and higher concentrations of nutrients.

The average American now consumes an average of 70 kilograms of sugar each year, most of it hidden away in soft drinks, cookies, cakes and processed snack foods. Thus, sugar, although not proven to be 'bad' by itself, provides only 'empty calories' that may lead to marginal nutritional deficiencies.

That the food choices of many Western people today might lead to marginal nutritional deficiencies is not surprising when you look at the following list of the most often purchased items in Australian supermarkets (Table 1). Supermarket surveys in the US show similar food choices, and year by year more and more Europeans and Asians are making similar food choices.

The only actual 'food items' are Heinz baked beans and Double Circle beetroot at 12. and 13., Kellogg's corn flakes (a highly processed food) at 19. and three brands of hydrogenated vegetable margarines (loaded with unhealthy trans-fatty acids) at 11., 23. and 25.

Table 1 List of the top 25 items actually purchased in supermarkets in Australia

1. Coca-Cola, 375 ml	14. Diet Coke, 1 litre
2. Coca-Cola, 1 litre	15. Bushell's tea
3. Coca-Cola, 2 litre	16. Cadbury Dairy Milk chocolate
4. Diet Coke, 375 ml	17. Pepsi Cola, 375 ml
5. Cherry Ripe	18. Coca-Cola, 1.5 litre
6. Nestlé's condensed milk	19. Kellogg's Corn Flakes
7. Tally Ho cigarette papers	20. Maggi 2-minute chicken noodles
8. Mars Bar	21. Generic brand lemon drink
9. Kit Kat	22. Panadol tablets, 24 pack
10. Crunchie Bar	23. Meadow Lea margarine
11. Eta 5-star margarine	24. Generic brand lemonade
12. Heinz baked beans	25. Mrs MacGregor's margarine
13. Double-Circle beetroot	

The other 19 items can all be considered 'Junk Food' high in sugars, fats, and various chemical preservatives, colouring and flavour enhancers, but low in both essential macro- and micro-nutrients required to maintain a healthy body and brain.

In 1970, Americans spent $6 billion on hamburgers, fries and other fast foods. By 2000, they were spending over $110 billion for fast food. Likewise, Kelly Brownell, author of *Food Fight*, found that in 1977 one in ten kids meals were from fast food restaurants, but by 1996 it was one meal in three. Americans now spend more money on fast food than on higher education, personal computers, computer software or new cars. From close inspection of the 1988–1994 Health and Nutrition Examination Survey, Dr. Alanna Moshfegh of the US department of Agriculture found that 'energy dense, non-nutrient poor' foods accounted for over 30% of American children's daily energy intake, with sweeteners and desserts together accounting for nearly 25% of the total.

REASONS FOR NUTRITIONAL DEFICIENCIES

A recent survey 4,700 Americans discovered that 25% of the food eaten in the past 24 hours consisted of 'empty calories', mostly 'energy dense' burgers, fries, and pizzas, while salty snacks and fruit juices made up an additional 5 per cent. Soda made up for 7.1 per cent of all the calories consumed, while nutritious foods such as fruits, nuts and vegetables accounted for only 10 per cent of the total calories.

What might drive this amazing increase in the consumption of Junk Food? Just consider the following statistics:

- The food industry has a $30 billion annual advertising budget.
- McDonald's reportedly spent $500 million on one ad campaign, while the National Cancer Institute spends about $1 million a year to promote eating five daily servings of fruits and vegetables.
- McDonald's and Coca-Cola each spent over $1.7 billion in advertising in 2004.
- Junk food advertisements now target children as young as 3 years old.
- Junk food marketers spent an estimated $15 billion in 2002 solely on marketing aimed at children, note that this is half of all the dollars spent on food advertising!

In 2003 the Food Standards Agency in the UK published a systematic review of the literature and found that, contrary to the claims of the food industry, ads for say, a particular chocolate bar not only boosted sales of that brand but also increased the overall consumption of chocolate bars. Likewise, a detailed study found that the more food ads children saw, the more calories they ate, and another study found that exposing children to TV ads for sweets made them more likely to choose a sweet over a piece of fruit when given the choice.

> **Key Concept: Junk & Fast Foods are low in Nutrients and Fibre, but high in Fats, Sugar and Salt and generally contain a number of Chemical Preservatives, Flavour Enhancers, and Artificial Colours and/or Sweeteners**
>
> Yet Junk and Fast Foods now make up a significant portion of the calories consumed, especially by children, adolescents and young adults in western societies. While energy-dense, these foods are nutrient poor. This may lead to Marginal Nutritional Deficiencies that compromise both their Health, and Brain Function. For many children, the chemical preservatives, flavour enhancers, and artificial colours also create their own problems from asthma to hyperactivity!

Another recent study reported in the Journal of the American Dietetic Association found that children as young as 2 years of age may be influenced in their food choices by a 30-second advertisement they see on television. The investigators found that a group of children aged 2 to 6 years were more likely to choose food products that they saw advertised compared with youngsters who had not viewed the commercials. In fact, just one or two commercials were enough to influence their choices.

This level of advertising for foods that are known to be deficit in nutrients certainly promotes food choices that may well lead to marginal nutritional deficiencies.

Part of the problem is truly the *excess of food* in the developed world, something totally new in human history. Fast Foods have been processed to increase their energy density at

REASONS FOR NUTRITIONAL DEFICIENCIES

the cost of nutrients by the replacement of nutrient-rich more expensive components with cheaper high-energy sugars and fats. Fast Foods really do 'Trick' the body, as their high energy density challenges the human appetite control systems with conditions for which they were never designed. Andrew Prentice of the International Nutrition Group at the London School of Hygiene and Tropical Medicine allowed groups of volunteers to eat as much as they liked. Those people given food with low energy density lost weight, while those people given food with high energy density put on as much as 65 grams of extra fat a day.

When he and Susan Jebb surveyed the energy-density of fast foods, he said 'I was astonished to see the energy density of a typical burger, which weighed at about 1200 kilojoules per 100 grams of food!' The typical energy density of the British diet is about 650 kj/100 g, and our bodies probably evolved to deal with an energy density of just 450 kj/100 g, typical of more traditional hunter-gather diets. Prentice and Jebb calculate that if someone over-consumes twice a week by 200 grams of Fast Food – about one burger – they could add up to 8 kilograms of body fat a year.

The result of this over-consumption of 'nutrient poor' but 'energy dense' fast and junk foods is obvious from an international survey of rates of obesity and overweight in 30,000 teenagers living in 14 industrialised countries conducted in 1997 and 1998. Among American teenagers, 15% of the girls and 14% of the boys were obese, and 31% of the girls and 28% of the boys were overweight. The definition of obese is being more than 33% above the normal weight of someone of your build and size. Clearly overweight is then weighing more than normal weight, but less than 33% above your normal rate.

The country with the lowest obesity rates was Lithuania. Among Lithuanian 15-year olds about 2% of the girls and 0.8% of the boys were obese, and only 8% of the girls and 5% of the boys were overweight. This is probably because at the time of the survey, 1998, Lithuania had few fast-food restaurants and relatively little processed foods in the markets, and Lithuanian teens had far less money to buy junk food snacks and fast food even where it was available. It would be interesting to repeat the survey now, as the number of fast food restaurants and availability of junk food snacks and processed foods has increased dramatically as Lithuania has prospered.

One of the primary sources of excess calories in many people's diet today is soft drinks, accounting for up to 10% of their total caloric intake. Indeed, soda and other soft drinks are not only sweet, but sweetened with High-Fructose Corn Syrup or HFCS, and in fact this is the only sweetener used in soft drinks in the United States. Remember that 20 pounds (almost 10 kilos) of corn syrup now consumed by every American each year, well this represents a >1000% increase in consumption of HFCS between 1967 and 2000, and it continues to rise. As of 2004, Americans over 2 years of age consumed on average 132 Kcal of HFCS per day, and the top 20% of consumers of sweeteners ingest 316 kcal per day of HFCS, basically an extra Big Mac a day.

Consumption of HFCS is now associated not only with the obesity epidemic sweeping the industriailised world, but also with the epidemic of type-2 diabetes, which was called adult on-set diabetes, but is now becoming increasingly common in children even under 10 years of age! HFCS is associated with obesity because absorption and metabolism of fructose differs from glucose. Unlike glucose, fructose does

> **Key Concept: Junk & Fast Foods have an extremely high 'Energy Density' but very low 'Nutrient Density', and it is this imbalance that results in Overweight & Obesity**
>
> Because the traditional diets that Man evolved with had a low 'energy density', and food was always scarce at least at some time of the year, the human appetite control systems were adapted to eating as much as you can when food was available, not to eat less than you can. Also foods high in nutrients tend to self-regulate the hunger system, resulting in eating less, as you have all the nutrients you need in what you just ate! Nutrient deficiency can literally drive people to eat 'dirt' in an attempt to get the minerals needed for survival, and likewise, lack of nutrients in Junk Foods often create addictive eating patterns. You eat Junk Food which is low in nutrients, so your body says – 'Eat more Food!' But if you eat more Junk Food, it just repeats the cycle, and the extremely high energy density of Junk Food just turns to fat!

not stimulate insulin release or increase leptin production. Leptin is a molecule produced by fat cells in the body which together with insulin release act as key signals in the regulation of food intake and body weight, and thus HFCS consumption is associated with increased energy intake and weight gain.

Likewise, a study of nutrient consumption and type-2 diabetes between 1909 and 1997 showed a strong correlation between the increase in sugar consumption, especially corn syrup, and decreased fibre intake with the increase in adult onset type-2 diabetes. Since 1997 the dominant form of corn syrup consumed is HFCS, and it appears this may be a major contributor to type-2 diabetes in children, something unheard of less than ten years ago in the United States, and in countries like Australia, less than five years ago.

GENETICS: THE GENES YOU INHERITED

Consuming *and* absorbing the RDA or DV of a nutrient is not actually the same thing. From the Western medical perspective, if you eat a balanced diet containing 100% of the RDA for all nutrients, you will be healthy. However, because each of us possesses different genes, we vary considerably in our ability to absorb nutrients from the food we eat, and if nutrients are not absorbed into the bloodstream, our cells cannot use them!

Even once absorbed into the blood, however, the nutrients then must be assimilated by being transported across the cell membrane, before they can be utilized within the cell to play their respective roles. Both absorption and assimilation of nutrients, however, is highly dependent upon having effective receptor and transporter molecules.

Thus, a major factor, often over-looked, is the genes you inherit from your parents and ancestors. For proper levels of nutrients to be absorbed, transporter and receptors molecules must be able to efficiently take up these nutrients into the blood from the gut and then from the blood into cells. And then once inside the cells, they must be efficiently converted into functional molecules needed for proper function. These receptor and transporter molecules are produced from the biochemical blueprint contained in your genes.

Genes are the particular sequence of DNA base pairs that encode the exact sequence of amino acids to make a specific protein, and slightly different versions of the same gene are produced by mutation. Therefore, the great majority of proteins are produced by many different genes, each with a slightly different amino acid sequence. These, different genes make slightly different versions of the same protein, our enzyme, transporter or receptor molecules, are called Alleles. Because the sequence of amino acids determines the shape of a protein, each Allele produces a slightly different shaped version of the protein, and in chemistry, the shape determines its function.

For many structural proteins these different shapes may have a negligible effect on their function. But for transporter, receptor and enzyme molecules – shape is all-important. These molecules have special shapes that allow them to interact with other specific molecules called the 'active site'. And shape of the molecule controls the effectiveness of the 'active site' that interacts with these other molecules. This in turn determines how effectively the molecule can do its job.

You can think of an 'active site' as the lock into which the key must fit for it to do its function, opening the lock. So if we have a 'receptor site keyhole', like a receptor molecule on the cell membrane, that is coded for by a faulty Allele, it just will not 'fit' the molecular key, say a hormone, causing the function of that hormone to be disrupted because the 'fit' is less functional. Have you ever struggled with a poorly copied key that sticks in the lock?

Likewise, if the molecular key, a precursor molecule such as an amino acid or fatty acid, does not fit the keyhole of the 'active site' on an enzyme very well, it cannot easily take part in enzymatic reactions. Thus even in the presence of lots of precursor molecules, the 'active site' on the enzyme coded for by a faulty Allele may slow down the whole reaction, and less neurotransmitter or cell membrane will be produced.

This reduction in neurotransmitter production may well result in a loss of integration brain function when you are stressed. You simply run out of enough neurotransmitter(s) to maintain function when high levels are needed. This may become apparent by such actions as suddenly forgetting what you were going to say, or just not being able to get your thoughts together.

The altered shapes of some Alleles may have no effect or only a minor effect on the activity of the receptor, transporter or enzyme they code for if located in a stable region well away from the 'active site'. However, only minor changes in shape near the 'active site' may have profound affect on receptor, transporter or enzyme efficiency and function.

For example, the Allele 1 for a Zinc Transporter may be very efficient and 'grab' 8 out of 10 zinc atoms and bring them into the blood or cell, whereas Allele 2 for the same Zinc Transporter may be very inefficient and only 'grab' 2 out of 10 zinc atoms.

Clearly, if you inherited Allele 2, you will have a difficult time absorbing or assimilating enough zinc atoms out of even a zinc rich diet to prevent becoming at least marginally deficiency of zinc. Choosing to eat highly processed foods low in zinc can only exacerbate your genetic tendency towards zinc deficiency. A deficiency of zinc can cause lethargy, poor concentration, memory and mood swings.

Differences in efficiency of various receptor, transporter

and enzyme molecules due to inheriting different Alleles is one the common mechanisms underlying marginal nutritional deficiencies for many people. Depending upon 'How?' ineffective the Allele you inherited is, you may have only minor deficiencies that may be treated by merely increasing the amount of zinc rich foods in your diet, or you may have major deficiencies that can only be corrected by supplementation.

> **Key Concept: Genes are the particular sequence of DNA base pairs that encode for the exact sequence of amino acids to make a specific protein**
>
> For the great majority of proteins there is not just one gene for each protein, but due to genetic mutations, many different genes for the same protein in the population. Each gene makes a slightly different amino acid sequence, thus different variations of the same enzyme, transporter or receptor molecules.
>
> Genes that code for the same protein, but differ in the exact sequence of amino acids are called *Alleles*. There may be 5 to as many as 40 different Alleles coding for each protein in the human body. Because the sequence of amino acids determines the shape of the protein, each Allele produces a slightly different shaped version of the protein, and in biochemistry – Shape determines Function. Thus, some Alleles produce fast, efficient enzymes, transporters or receptors, and others slower, less efficient versions.

Clearly, one way to compensate for this genetic defect is to simply to increase levels of precursor molecules that then drive the slower enzyme, transporter or receptor molecules faster to re-establish normal function. When the defect is slight, merely increasing foods rich in the specific precursor molecules may be sufficient, but if the defect is more severe, you may need to do this by supplementation.

By providing increased levels of nutrients and precursor molecules these slow reactions are driven faster to produce normal levels of nutrients in the cells, and therefore, normal levels of enzyme activity. This normalized enzyme activity is then able to produce the levels of neurotransmitters required to maintain effective integrated brain function, permitting optimum mental performance.

> **Key Concept: Normal Genetic variation results in some people having inefficient Enzymes, Receptors, and Transporter Molecules that may result in Marginal Nutritional Deficiencies even when eating a 'Balanced' Diet**
>
> Due to the presence of several Alleles coding for the same enzyme, receptor or transporter molecule, some people have fast, efficient enzymes, receptors or transporters, while other people may have inherited a faulty slower version of these important molecules. These genetic defects may lead to marginal nutritional deficiencies even when eating a 'balanced diet'. The presence of very inefficient or slow enzymes, receptors or transporters may need to be addressed by nutrient supplementation.

Food Choices and Diet: What to do?

Clearly people in the developed Western world live in a time of food abundance and choice unknown before the past 30 years. There is an abundance not only of food, but of food choices! When my generation was young, while there was far more food available of a much wider variety than our parent's generation, most of this food was still grown in time-honoured ways without genetic modification, and the massive amounts of chemical fertilizers, pesticides, ripening agents, etc. with which food is treated today. Also our food choices were relative to today, very limited, as there was very little Junk Food available and there were basically five types of soft drinks and a half a dozen of so types of candy bars, and these were restricted to specific shelves in the supermarkets and general stores (there were no convenience stores), not pushed in your face everywhere you go as today!

Even when you know you are making less than optimal food choices, how can you change your diet? And which diet do you choose, as there are now a plethora of diet plans, and every magazine suggests a new diet every week? My suggestion is to look for a diet plan that 'fits' you and your physiology the best, as we all vary genetically, and what is perfect for one person is a disaster for another. In Section 2 of the Annotated Suggested Readings at the back of the book I discuss two quite different, yet very successful and effective Diet Programs that support optimal health and wellbeing, and you are referred to these resources for further information on this important topic.

Chapter 3

NUTRITION AND HOW IT WORKS

INTRODUCTION TO NUTRIENTS

Nutrition is the amino acids that form the proteins for our bodies, the glucose that generates the energy in these cells, the fatty acids that make up the membranes of our cells, and the vitamins and minerals needed to maintain body function.

These nutrients can be divided into two basic categories:– macro-nutrients and micro-nutrients.

Macro-nutrients: These are the common constituents of our bodies such as protein, fats, calcium and phosphorus.

Micro-nutrients: These are the nutrients that are present in only small concentrations such as vitamins, minerals (such as iron, zinc and copper) and the 'trace elements' present in even smaller concentrations. While most people may not be familiar with these micro-nutrients, they are essential in maintaining our health and optimising our brain function.

Let's start with the **Macro-nutrients** since they are the major building blocks of our body. There are basically five different kinds:

Proteins – these are made up of chains of amino acids and form the fabric of our muscles, tendons and ligaments.

Fatty acids – form the fabric of our cell membranes, hormones and fat deposits.

Bone Minerals – these are Calcium and Phosphorus that make up our bones and,

Body Minerals – Potassium, Sulfur, Sodium, Chloride, Calcium, Magnesium, Iodide and Iron that keep our nerves firing, our muscles contracting and our blood and hormones working. Together with bone minerals they make up about 4% of the total body weight.

Carbohydrates – the sugar and starch that provide the energy to run our bodies and brain.

Figure 3a (at the end of this chapter) lists the major macro-nutrients and summarises their role in the human body.

The **Micro-nutrients** play a more subtle but equally important role in our health. The most common ones important to health fall into two groups:

Vitamins – organic compounds that help regulate our bodily activities known by letters, and sometimes followed by numbers, E.g. Vitamin A, Vitamin C, and Vitamin B_1.

Micro-minerals and Trace Elements – micro-minerals are present in only small concentrations but essential for our metabolism and essential for energy transactions in the body. Examples of micro-minerals are manganese, copper, and zinc. Trace elements are so called because only trace amounts of them are needed. Examples are: chromium, molybdenum and selenium.

> ### Key Concept: Macro- and Micro-Nutrients
>
> *Macro-Nutrients* are the common constituents of our bodies such as proteins, fats, carbohydrates and minerals such as calcium magnesium and phosphorus of our bones.
>
> *Micro-Nutrients* are the nutrients present only in small concentrations – vitamins, minerals (such as iron, zinc and copper) and the 'trace elements' present in even smaller concentrations.

THE ROLE OF MACRO-NUTRIENTS IN THE BODY AND BRAIN

There is a key nutritional concept, the importance of which will be fully understood as we progress. There are nutrients we make in our bodies from other nutrients we have eaten, and there are those that we have to get directly from our diet that, once consumed, can only be replaced by eating more of that type of nutrient. The nutrients that cannot be made by the body are called 'essential' nutrients, while those that can be generated from 'essential' nutrients we have eaten are called 'non-essential' nutrients. Amino acids are an example of this key concept so we can talk of 'essential' and 'non-essential' amino acids.

Since the bulk of our body is made up of protein in the form of muscle and other connective tissue, and proteins are simply long chains of amino acids, then amino acids must be necessary nutrients for a healthy body. However, our bodies can easily produce only 8 out of the 20 amino acids we need. A further 2 amino acids are made in insufficient quantities for practical purposes, so these, along with the remaining 10, have to be obtained directly from food. Therefore, the primary source for amino acids is foods rich in these essential nutrients that, fortunately, are not too difficult to obtain since meat and most vegetables have high levels of amino acids.

The function of proteins is largely controlled by what are called 'functional groups' that are attached to the chain of amino acids making up each specific protein. These functional groups may be attached to the beginning or end of the protein or may occur as 'side-chains' sprouting off the middle of the chain. The most common 'functional groups' are: the Hydroxyl group, denoted chemically as $-OH$; the amine group, denoted as $-NH_3$; the methyl group, denoted as $-CH_3$, and the aromatic ring, denoted as a six-sided hexagram of carbon-atoms with a ring of electrons in the middle, e.g. ⬡. What makes these 'functional groups' so important is that they change the shape of the protein, and hence its function.

Amino acids are not just important in producing body tissue. There is a small group of non-essential and essential amino acids that are precursors to important brain neurotransmitters controlling nerve function. These neurotransmitters are molecules that act as chemical messengers by carrying a nerve impulse (message) from one nerve cell to another. This group of essential dietary amino acids includes:

- *Tyrosine*, the precursor of both Dopamine and Noradrenalin; two important neurotransmitters that control how 'good' we feel, and how well we handle stress.
- *Phenylalanine*, the precursor of Dopamine; a 'key' neurotransmitter involved in the reward and movement systems of the brain.

- *Tryptophan*, the only precursor of Serotonin; the calming neurotransmitter involved in not only vanquishing depression, but also synchronising brain function.
- *Glutamine*, the precursor of Glutamate; the major excitatory neurotransmitter in the brain 'turning on' brain activity;
- *Gamma Amino Butyric Acid* (GABA) and *Glycine*, the primary inhibitory neurotransmitters in the brain 'turning off' brain activity.

> **Key Concept: Essential and Non-Essential Nutrients:**
>
> **(E.g. There are Essential Amino Acids and Essential Fatty Acids)**
>
> Many nutrients we make in our bodies from other nutrients we have eaten, but there are some nutrients that cannot be made by the body and must be taken in directly from our diet. These are called *Essential Nutrients*, while those that can be manufactured from the 'essential' nutrients we have eaten are called *Non-essential Nutrients*.

Fatty acids: like amino acids, may also form long chains and come in 'essential' and 'non-essential' forms. Fatty acids are important as structural components of cell membranes in the body, which if spread out flat would cover 17 football fields. Fatty acids are also the precursors to important sterol hormones such as estrogen, testosterone and cortisol.

Each cell is covered by a lipid (fatty acid) bilayer that holds all of the cellular contents inside and provides the structure over which all nutrients and chemicals must pass to enter the cell. Much like the wall of a medieval city, the cell membrane has portals of various sizes for a variety of functions. Some of these are tubes or tunnels made of protein molecules that go through the cell membrane (like the gates in the ancient walls), while others are protein molecules protruding through the membrane that 'grab' specific passing molecules and spirit them into the cell, like the secret passages in the city walls only accessible with the 'password' known to the guard.

In the same way that the make-up of the wall determines how it functions (cemented hard granite or loose soft sandstone), the type of fatty acids making up the membrane are also very important, because they control the flexibility and hence the function of the membrane. While the fatty acid membrane walls of most cells in the body are quite similar in structure, the neuronal membranes of the brain have a unique fatty acid composition giving them their 'special' properties, like the walls of the inner sanctum of the fortress.

Like amino acids, fatty acids may also have various 'functional groups' attached to them that change their activity or function. Most people today are aware of cholesterol and both the 'good' high-density lipo(fatty acid)-proteins (HDLS) and the 'bad' low-density lipo-proteins (LDLs) in the blood. The high-density lipo-proteins are important transport molecules in the blood, whereas the low-density lipo-proteins are considered bad because they are implicated in making the plaques on arterial walls leading to arteriosclerosis.

Other important fatty acid functional groups act as receptors on the membrane surface to activate functions within the cell. Therefore, when a hormone attaches itself to a membrane receptor, it activates certain functions within the cell relevant to the function of that hormone. In fact, this turning on of internal cellular action by hormone-receptor activation is the basis for all hormonal action. It is also why only specific cells or glands are affected by hormones – only those cells that have the 'receptor' molecule for a hormone can respond to the presence of this hormone.

> ### Key Concept: Functional Groups
>
> **Many proteins and fatty acids have 'functional groups' that attach to the primary amino acid or fatty acid chain.** Because each functional group has a different shape and activity, they give the protein or fatty acid a different function compared to the same protein or fatty acid without this group attached – hence the name, 'functional group'. Examples are the hydroxyl group, the amine group, the methyl group and the aromatic ring, or benzene ring.

Although generally regarded as 'bad' by the public, one of the most important fatty acids is cholesterol. Cholesterol is the precursor of most of the sterol hormones that control our sexuality, such as Testosterone, Estrogen and Progesterone, as well as the important hormones controlling our immune system like Cortisone and Cortisol.

Macro-minerals are minerals present in such high concentrations in the body that they are also classified as macronutrients. Examples of these are sodium, potassium, chlorine, calcium, magnesium and phosphorus. These macro-minerals play a major role in our structure and function. The skeleton is largely composed of calcium and phosphorus, while all nerve conduction is dependent upon the flux of potassium and sodium across the neuronal membrane. Magnesium and calcium play major roles in muscle contraction (the reason magnesium deficiency may result in muscle cramps) and the control of nerve function.

Carbohydrates, like sugar and starch, are other important macro-nutrients as they provide the building blocks for functional components (like the polysaccharides in mucus protecting the stomach wall and lubricating the intestines) and the energy from glucose necessary to power our body and brain.

Clearly, when we cannot provide our bodies with enough of these macro-nutrients, we suffer health problems (including starvation) but just having enough or even an excess of these macro-nutrients is not enough to make us healthy. Why is this so?

Figure 3b (at the end of this chapter) lists the major micro-nutrients, sources of each type in our diet and summarises their role in the human body.

Micro-nutrients in the Body

The micro-nutrients, although less obvious, play an equally important role in our health, as they often provide critical functions such as enzyme activation, or activation of membrane channels controlling nerve and cellular function. They may also be the 'key' elements in many important molecules like chlorophyll and hemoglobin. The micro-nutrients are not necessarily 'small', just present in very small amounts compared to the macro-nutrients that are present in relatively large amounts in the body.

Figure 3b (at the end of this chapter) lists the major micro-nutrients, sources of each type in our diet and summarises their role in the human body.

THE ROLE OF MICRO-NUTRIENTS IN THE BODY

While Micro-nutrients are present in only tiny concentrations, their primary function is regulatory, to control the interactions and functions of the macro-nutrients and the structures made up of these macro-nutrients. So while chromium is only needed in micro-gram (that's millionths of a gram) concentration, deficiency of chromium may have profound effects on heart and bodily functions, especially the metabolism of glucose. This is because the chromium acts as the linchpin holding together the two molecules of the Glucose-Tolerance Factor necessary for proper glucose metabolism. This is especially important for the brain that uses more glucose gram-for-gram than any other tissue in the body.

The micro-nutrients also play a major role in the rate of enzymatic reactions, as they are often the controlling factors that enable or modulate enzyme function. One important class of micro-minerals is the *co-factors* that along with vitamins control enzyme action. To prevent enzymes from being active before they are in the correct location, enzymes are made in an inactive form called an *apoenzyme*. To become the active functional enzyme, the apoenzyme must first join with a micro-mineral co-factor such as copper, iron or zinc or a *coenzyme* such as vitamin B_6. This changes the shape of the apoenzyme making it into an active enzyme. (See Fig. 2.)

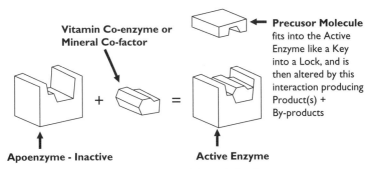

Figure 2 Vitamin Co-enzymes or Mineral Co-factors: are necessary in order to convert the inactive Apoenzyme into the Active Enzyme so that it can interact with the Precursor Molecule.

Clearly, deficiency of either vitamin co-enzymes or mineral co-factors will significantly reduce enzyme activity, a major reason people feel so tired when they are deficient in B-vitamins that are essential to activate the enzymes needed to turn glucose into Adenosine Triphosphate (ATP) the energy currency of the body.

Since the biochemistry of the body is very complex, like a factory making complex machinery, slowing down one enzyme due to deficiency of a required co-factor or coenzyme may slow down the whole output of finished machinery. If only one part is in short supply due to one slow machine making a vital part, the production of the whole factory is slowed down. This is one reason why is not sufficient to just follow the lists of RDA's as such a wide variety of nutrients is required, and people vary so much in their genetic constitution, that one slow enzyme may compromise whole brain functions. (Discussed in detail in Chapter 4.)

Since enzymes are the catalysts that determine how fast parts of the body are made or metabolised, mineral co-factor or vitamin deficiency can disrupt production of molecules important to healthy function, as well as the elimination of toxins and other molecules important to healthy function. It must be realised that when two molecules are made into a more complex molecule, normally each of the original molecules also had to be made by other enzymes. So the body needs the 'correct' nutrient matrix to create synergy between the different nutrients, with the function of one nutrient supporting the function of the other. This nutrient matrix specifies both the types and quantities of the various nutrients required. There may be a chain of several steps required to create the more complex molecules, with each step requiring a different enzyme, and each enzyme is coded for by a specific gene.

Without adequate levels of micro-nutrients necessary to maintain efficient production of these precursor molecules,

> **Key Concept: Enzymes need a Co-Enzyme or a Co-Factor to become Active**
>
> **Vitamins act as Co-enzymes & Minerals act as Co-Factors**
>
> **One of the major roles of micro-nutrients is to control enzyme action in the body.** To prevent enzymes from being active before they are in the correct location, they are made in an inactive form called an *apoenzyme*. To become the active functional enzyme, the apoenzyme must first join with a micro-mineral *co-factor* such as copper or zinc or a *coenzyme* such as vitamin B_6. This changes the shape of the apoenzyme creating an *Active Enzyme.*

the whole process may also slow down. Thus, the entire production line slows down because only one factory supplying a vital part has too many workers out sick or working too slowly.

Figures 3a & 3b on the following pages summarise the sources and functions of both Macro-nutrients and Micro-nutrients of the Human Body. They are basically an outline for much of the discussion in many of the chapters that follow. Appendices 1 & 2 summarise information on the Vitamins, Minerals and Trace elements, their sources, Recommended Dietary Allowances (RDAs) and Therapeutic Dosage Range (TDRs), functions and symptoms of deficiency.

NUTRITION FOR THE BRAIN

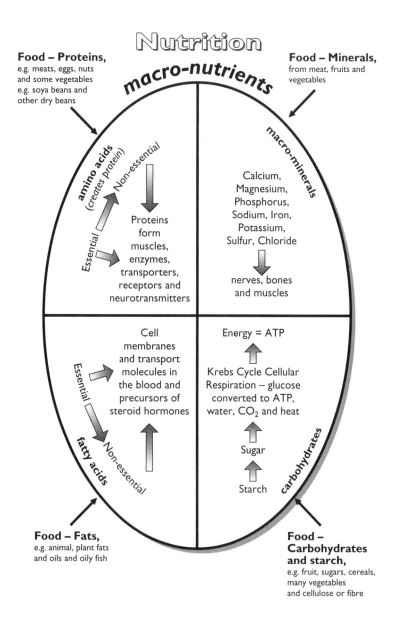

Figure 3a The Macro-Nutrients of the Body.

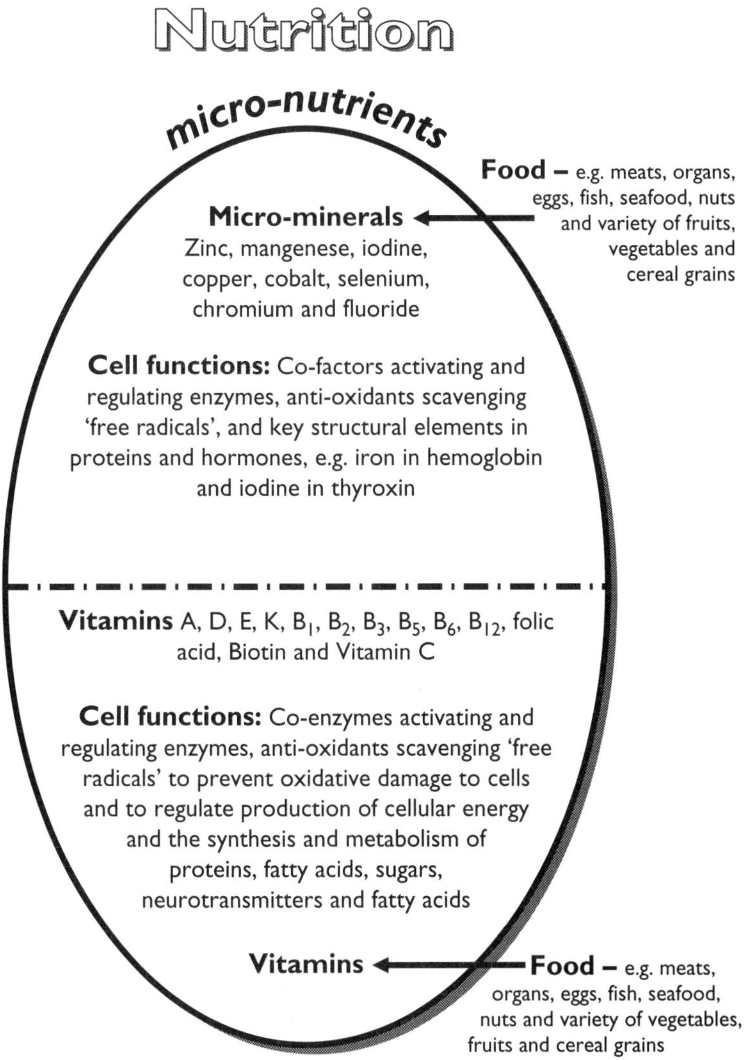

NUTRITION AND HOW IT WORKS

Nutrition

micro-nutrients

Micro-minerals ← **Food** – e.g. meats, organs, eggs, fish, seafood, nuts and variety of fruits, vegetables and cereal grains

Zinc, mangenese, iodine, copper, cobalt, selenium, chromium and fluoride

Cell functions: Co-factors activating and regulating enzymes, anti-oxidants scavenging 'free radicals', and key structural elements in proteins and hormones, e.g. iron in hemoglobin and iodine in thyroxin

---

Vitamins A, D, E, K, B_1, B_2, B_3, B_5, B_6, B_{12}, folic acid, Biotin and Vitamin C

Cell functions: Co-enzymes activating and regulating enzymes, anti-oxidants scavenging 'free radicals' to prevent oxidative damage to cells and to regulate production of cellular energy and the synthesis and metabolism of proteins, fatty acids, sugars, neurotransmitters and fatty acids

Vitamins ← **Food** – e.g. meats, organs, eggs, fish, seafood, nuts and variety of vegetables, fruits and cereal grains

Figure 3b The Micro-Nutrients of the Body

Chapter 4

THE ZINC CONNECTION

While any nutrient may be either marginally or absolutely deficient, marginal deficiencies are far more common. Zinc deficiency will be discussed in some detail below as an example of the effects of a single important micro-nutrient deficiency upon brain function, but much of this discussion also applies to most other nutrients required for optimum physical and mental performance, especially iron and copper whose biochemical dynamics are similar to zinc.

Did you know that zinc and iron are two of the most common micro-nutrient deficiencies in the world today with over two billion people affected by these deficiencies? In the Western developed world, both zinc and iron deficiency is widespread with more than half of the people having overt iron deficiency, and between 20% to perhaps as much as 50% also zinc deficient, although the extent of zinc deficiency has been less well studied. Interestingly, zinc deficiency in human populations has received relatively little study until very recently, but it has been widely studied for years in domestic animals where it is considered important in optimising dietary aspects of husbandry.

According to Dr. Khalsa in his book *Brain Longevity*,

almost 90% of Americans fail to meet even the so-called RDA for zinc on most days. He suggest that one reason for this is that you would have to eat almost 2500 calories per day to get enough zinc from common foods, unless you were carefully targeting eating foods rich in zinc.

The Dynamics of Zinc in the Body

Most people have little idea about what happens to the nutrients in their food during the process of digestion, and the processes that control and regulate nutrient use in the body – partly because this is a very complex subject obscured by scientific jargon. I will attempt to provide a succinct overview of these complex processes, so the reader can understand where it might all go wrong.

After chewing and swallowing our food, it goes to the stomach for the first stages of digestion, and then into the small intestine where digestion is finished and the nutrients are absorbed into the blood, transported to our cells, then assimilated into the cells where they are utilized in specific biochemical reactions that run our bodies and our brain. We will look at each of these important processes briefly in order to gain a functional understanding of what can go wrong to create these deficiencies.

Sources of Zinc Deficiency

As discussed in Chapter 2, the most overt source of zinc and other nutrient deficiency is having inadequate levels in your diet. For instance, the 1997 National Health and Nutrition Examination Survey found that 15 to 20% of Americans were overtly iron deficient. Zinc deficiency was widespread with 53 to 78% of females aged 12 to 79 having zinc intakes below the RDA, and all age and gender groups demonstrated a similar trend. Foodstuffs high in zinc include shellfish, steak, eggs, sunflower seeds, sesame seeds, pumpkin seeds, pecans, legumes and beans, most of which are not bountiful in the diets of children and adolescents today. While you should take in at least 15 mg of zinc a day, analyses of 'well-rounded' diets served at cafeterias and hospitals show only 8 to 11 mg per day is provided.

Lack of a nutrient like zinc in the diet is only one of a number of factors that can result in nutrient deficiencies even when, technically, adequate levels have been consumed. How can this be?

Digestion: Release of Nutrients from our Food

When you eat, you chew food to break it up into small pieces to increase the surface area for enzymatic reaction so the nutrients can be released from the food. This takes both adequate levels of the digestive enzyme that digests proteins and strong stomach acid. It is the acid in the stomach that converts the inactive pepsinogen (the apoenzyme) into pepsin the active enzyme that actively chews up the proteins of your steak. If you are hypochloric (have low levels of hydrochloric acid in your stomach) as is the case with up to 50% of Americans, then digestion and release of nutrients from food is compromised, and many nutrients may well pass out of the body before being digested to the point of release.

Absorption: Uptake of Nutrients from the digested Food

After the initial digestion in the stomach, the nutrients like zinc in your food are then released into the small intestine, generally attached to proteins (chelated) or in colloidal complexes that can then be taken up across the intestinal wall, a process called *Absorption*. However, even before absorption can occur there can be trouble brewing. First of all, incomplete digestion may not release all of the zinc you did eat, so for practical purposes, the actual levels of zinc that entered your blood from adequate levels in your food are just not enough to support optimal function.

Interaction with other Gut Molecules: Role of Nutrient Inhibitors

Secondly, even before the zinc can get to the gut wall to be absorbed, it must first 'fight' its way through a throng of sometimes hostile molecules, some of which are 'inhibitors' blocking zinc's way to the gut wall. One of the primary reasons for zinc, and also iron, deficiency when eating a diet with adequate levels of these micro-nutrients, is the presence of *Phytates* from cereal products, legumes and nuts that are often consumed at the same time as meat, the primary source of both zinc and iron, and zinc-containing vegetables. Phytates are molecules that bind firmly to zinc and, in a sense, grab the zinc by the arm and escort it directly out of your body.

Ingesting cereal products, that have high phytate levels, with foods, even those rich in zinc, may greatly reduce the bioavailability of zinc to your body. Fortunately, the fermentation of grains by yeast enzymes during the preparation of bread greatly reduces phytate levels. In contrast, the extrusion process used in the preparation of some breakfast cereals and many junk foods does not change phytate concentrations, and may in fact increase them. Indeed, populations relying heavily upon unleavened breads (with no yeast fermentation) and grain products often suffer from zinc deficiency in spite of eating other foods relatively rich in zinc.

And of course junk foods often contain food colouring like tartrazine (yellow #5), that also bind zinc strongly, and may remove zinc from the bloodstream as well as the gut.

Absorption and Assimilation: The Role of Transporters

The process of absorption for many of your nutrients, especially the larger molecules like zinc and iron, is dependent upon effective transporter molecules being in abundant supply, and reduced numbers of transporters or less efficient transporters means lower amounts of nutrients being absorbed.

(Remember, transporter molecules latch onto zinc atoms or zinc complexed to other molecules and spirit them across the cell membrane into the blood from the gut, then into the cell from the blood or into cell organelles from the intracellular fluid.)

So not only do transporter molecules play a vital role in getting zinc and other micro-nutrients into the blood from the gut, but they also control the movement of nutrients from the blood into the cells, a process called *Assimilation*, and even how and where they will be used within the cells, a

Table 2 Zinc Transporters & their Respective Functions

Zinc Transporter	Transporter Function
Zn T-1	Is the zinc transporter required for absorption from the intestine into the blood.
Zn T-2	Occurs only in the kidney and is responsible for urinary excretion of zinc.
Zn T-3	Is found exclusively in the brain and is the transporter primarily responsible for loading zinc into the brain.
Zn T-4	Is found only in breast cells and is responsible for delivering zinc into the milk.
Zn T-5	Is found in the pancreas and loads zinc into the digestive enzymes.
Zn T-6	Is ubiquitous being found in most cell types, and is responsible for zinc in the cytoplasm (the intracellular fluid of the cells) getting into specific cell organelles like the Mitochondria, the powerhouses of the cells.

process called *Utilization*. For example, there are 6 different types of zinc transporters, called Zn T-1 to Zn T-6.

ROLES THAT ZINC PLAYS IN THE BODY

Zinc is one of the most important minerals in the body because it is a co-factor that activates more than 300 enzymes and forms an important structural element of hundreds of structural proteins in the body, called Zinc-finger proteins, as well as playing an essential role in the function of metalloproteins such as metallothionein – a major free-radical scavenger and heavy metal detoxifying protein in the liver. Zinc, therefore plays many vital roles in our normal function. These roles include:

- Being major structural components, the Zinc-finger proteins.
- Being an important antioxidant.
- Supporting Immune System function & controlling Inflammation.
- Regulating neuron growth and myelination (insulation of neurons).
- Regulating neurotransmitter release.
- Regulating neurotransmitter metabolism.
- Regulating fatty acid metabolism.
- Regulating production of digestive enzyme.
- Regulating the production of detoxification enzymes and molecules in the liver.
- Regulating metabolism and transport of hormones.

To summarise some of these vital roles, zinc or molecules in which zinc plays a critical role like metallothioneins and superoxide dismutase (SOD) are powerful antioxidants scavenging free-radicals that damage the cells of the body, including DNA which may result in mutations leading to cancer. Zinc is also involved in many aspects of DNA structure and function, including DNA repair. It is also integral to the stability and expression of genes. Zinc plays a major role in both the production and regulation of both neurotransmitters and fatty acids essential for normal brain and body function, a topic that will be discussed in some detail below.

Zinc also plays some vital roles in immune system function, such as the maturation of T-cells, including Natural Killer Cell activity, and the cellular mechanisms maintaining homeostasis. Zinc also plays a major role in the control of inflammation, and Zinc-finger proteins protect the cell from cytotoxicity, including apoptosis or programmed cell death.

Zinc is vital as a component of enzymes necessary for major metabolic pathways, and is thus essential for life. The major synergistic nutrients, nutrients that enhance zinc's function: as a potent Antioxidant include Selenium, Magnesium, Manganese, Copper, Iron and the vitamin Pyridoxine (B_6); in neurotransmitter and fatty acid production include Magnesium, Selenium and vitamin C; in supporting immune system function, Copper, Magnesium and vitamin C. So while zinc is involved in many functions, it requires an ancillary nutrient matrix to enhance and maximize these functions.

These numerous roles that zinc plays in our health are not unexpected as zinc is the second most abundant micronutrient in the body after iron. If you subtract the amount of iron found in hemoglobin in the blood, then zinc is the most abundant trace metal found in the human body.

Therefore, chronic zinc deficiency from whatever source, deficiency in the diet, food choices or genetic make-up, may produce widespread symptoms within the body.

Signs and Symptoms of Zinc Deficiency

- Growth retardation
- Neuropathy
- Diarrhoea
- Dermatitis
- Immune dysfunction
- Impaired DNA synthesis & repair
- Impaired sense of smell & taste
- Eating disorders
- Cognitive impairment

- Memory disorders
- Dyscalculia (Difficulty understanding Arithmetic and Mathematics)
- Hyperactivity Attention Deficit Disorder (ADHD)
- Sleep disorders

Causes of Zinc Deficiency

Zinc deficiency, especially marginal deficiency, may result from low levels zinc in your diet even if your diet has sufficient or even excess calories, as in a diet rich in junk and fast foods. These deficiencies may also result from the binding of zinc by phytates in food you eat, especially some junk and fast foods, for behavioural reasons such as excessive alcohol intake and smoking, or for genetic reasons. Chapter 2 discussed the reasons many people's diet are deficient in key micronutrients like zinc.

When the zinc deficiency from any cause is also accompanied by deficiencies of other B-group Vitamins, further symptoms are likely to occur, as zinc deficiency may exacerbate or be exacerbated by the deficiency of other macro- and micro-nutrients. This is not surprising when you consider the number of biochemical reactions in which zinc plays a role, with each reaction requiring zinc plus other ancillary nutrients to proceed unimpeded. Table 3 below summarizes these additional symptoms.

This clearly demonstrates that deficiency of one micronutrient can be strongly exacerbated by the deficiency of other nutrients, because nutrients work together to optimize our functions. This has been seen in a number of studies where children with chronic zinc deficiency that were

Table 3 Zinc Deficiency in association with B-Group Vitamin Deficiencies

- Increased Tendency towards Infection
- Reduced or Slow Wound Healing
- Skin may Bruise easily
- Loss of Hair
- Trophic Disorders:
  - Nails break easily, are brittle
  - Hair is often straw-like and breaks easily
  - Skin often is cracked
  - Horizontal ridges on the nails
  - Poor Protein Digestion and medicine response
  - Nausea in the mornings
- White lines or flecks in the nails
- Pale skin and sensitivity to sunlight
- Often pain in the upper abdomen
- Irregular menses in women and impotence in men

supplemented with zinc alone often showed little improvement in growth, behaviour and mental performance. Yet children supplemented with a range of macro- and micronutrients plus zinc showed major changes in growth, behaviour and cognitive abilities.

Genetic Zinc Deficiency: A Little Considered Cause of Deficiency

Remember that genes code for proteins, for instance the Zn T-1 to ZN T-6 transporters, and that there are a number of different Alleles that code for each of these transporters. So people with an Allele that makes a 'perfect' Zn T-1 molecule, will rapidly absorb most of the zinc in their food. In contrast, people with an Allele that makes a less efficient Zn T-1 molecule, will only be able to absorb a small amount of the zinc in their food.

It follows therefore, that some people can be genetically more capable of absorbing, assimilating and utilising the zinc in their food than others. So if you happen to have an Allele that makes an inefficient zinc transporter, you may well end up with at least a marginal zinc deficiency even when eating a diet rich in zinc.

While Kryptopyrroluria is a known genetic mutation affecting zinc dynamics in the brain and body discussed below, there are very likely many other less severe mutations in genes regulating zinc dynamics that are as yet not recognised, but may still result in chronic zinc deficiency.

KRYPTOPYRROLURIA: A GENETIC MUTATION CREATING A COMBINED ZINC AND B_6 DEFICIENCY

One genetic based cause of zinc deficiency is called Kryptopyrroluria, also known as Mal varia or Mauve Factor. Kryptopyrroluria results in chronic widespread zinc and B_6 deficiency throughout the body and brain. The medical fraternity largely ignored it until very recently, even though this condition was first described over 30 years ago and a number of scientific papers have been published on this topic by Dr. Carl C. Pfeiffer among others.

Kryptopyrrole was the initial molecule found in the urine of people with this genetic fault, and then ascribed as the causative agent of this condition. Later research demonstrated that it was not Kryptopyrrole, but rather a very similar molecule, hemopyrrole (chemically Dihydroxy-hemopyrrolene-5-one) that was actually the agent involved in kryptopyrroluria. Both substances respond to the test for the metabolic products in the urine, producing a mauve coloured spot – the reason the condition was at first called the Mauve Factor. Because Kryptopyrrole was first designated the causative agent and the condition was called kryptopyrroluria, I will continue to use this word throughout the book, even though this is not exactly correct chemically.

During the normal break down of the haem molecule, a major component of Haemoglobin, the porphyrin molecule is broken down into a metabolite that is converted to a water-soluble form by conjugation and eliminated in the urine. However, when there is a faulty gene coding for the liver enzyme converting the porphyrin molecule to the water-soluble molecule, haemopyrrole is produced instead and excreted in the urine. Since pyrroles are not normally excreted in urine, a chemical test for pyrroles is used to detect this condition.

Unfortunately, kryptopyrrole tends to form a conjugate with vitamin B_6, forming a kryptopyrrole–B_6 complex. This chemical process is called *conjugation*, from conjugal – to be married. Just like getting married, the kryptopyrrole molecule grabs the Vitamin B_6 by the hand and the combination is then excreted together, like the groom takes the brides hand and leads her from the church.

Not only is the kryptopyrrole–B_6 complex excreted in the urine, but this complex then *chelates* a zinc atom and removes zinc as well as vitamin B_6 from the body. The net result is that for people with a faulty version of the gene resulting in kryptopyrrole production, excrete not only the kryptopyrrole–B_6 complex, but also the chelated zinc. This may lead to widespread zinc and B_6 deficiency throughout the body and brain.

To *chelate* or *chelaton* come from the Greek for pincer, as the kryptopyrrole–B_6 complex forms a pincer that grasps the zinc atom, which is then excreted along with this complex in the urine. (See Fig. 4.)

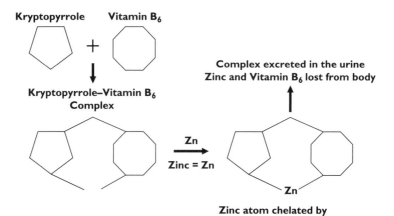

Figure 4 The Conjugation of Kryptopyrrole to Vitamin B_6 and then Chelation of Zinc Eliminates both Zinc and Vitamin B_6 from the Body resulting in Chronic Zinc and Vitamin B_6 Deficiency.

Thus, for every molecule of haem that is metabolized, people with this genetic fault lose an atom of zinc and a molecule of B_6. Since you are metabolizing haem molecules every day all day long, these people may become progressively zinc and B_6 deficient, even eating a zinc rich diet containing normal levels of vitamin B_6. Because of this constant extraordinary loss of these nutrients, even taking normal zinc and B_6 supplements may not resolve this problem.

It is estimated that more than 10% of the population have an enzymatic disturbance in haem metabolism, but most of these enzyme faults do not lead to overt problems, rather they are compensated by other biochemical mechanisms – the reason this condition was called Krypto-(hidden)-pyrroluria. So you do not have a single gene causing this problem, but rather a group of different Alleles coding for slightly different versions of the pyrrole metabolising enzymes.

Therefore this condition varies over a range of degrees of expression from mild to severe, with varying degrees of expression of the symptoms given in Table 4. Interestingly, studies have shown that only 11% of normal people shown detectable levels of pyrroles in their urine, in contrast to 24% of 'disturbed' children, 42% of psychiatric patients, and 52% of schizophrenic patients have moderate to high levels of pyrroles in their urine.

It is important to realize that genes are *not destiny*, but *only a propensity* to express certain aspects of metabolism or behaviour. The critical factor that decompensates the expression of kryptopyrroluria is stress! Therefore in the absence of high stress levels, people with kryptopyrroluria may not express any or only mildly the symptoms in Table 4, but under stress their compensatory mechanisms breakdown and they may then overtly express a variety of these symptoms. Indeed, one of the reasons that this condition has not been embraced by the medical community for its likely role in the conditions given in Table 4 is exactly because they are so diverse and also so variable in expression both between people and within the same person over time.

A brief inspection of Table 4 overleaf listing the symptoms

of zinc *and* vitamin B_6 deficiency resulting from Kryptopyrroluria clearly demonstrate the vital role deficiency of these two micronutrients play in body and brain function.

Table 4 Clinical Symptoms Associated with Chronic Genetic Zinc Deficiency resulting from Kryptopyrroluria

- Various Neurological symptoms not associated with known neurological disease
- Cognitive (Thinking) Disorders
- Hyperactivity Attention Deficit Disorder (ADHD) Syndrome
- Perceptual Disorders
- Memory disorders of unclear origin
- Often Bad Short-term Memory and especially Name-Memory
- Often poor Handwriting
- Psychological Disorders:
  – Borderline Type Personality between Genius and Craziness
  – Neurotic Anxiety in Stress Situations
  – Depression
  – Schizophrenia-like symptoms, often mis-diagnosed as Schizophrenia
  – Strong Emotional Ups & Downs
- Nervous Exhaustion
- Sleep Disturbances & Disorders
- Only partial Dream Memory
- Symptoms other than Neuropsychological Symptoms:
  – Rheumatological complaints
  – Unspecified Autoimmune Phenomena

So, if you have Kryptopyrroluria or less than ideal Alleles coding for your zinc transporters or receptors, are you then doomed to a lifetime of some or all of the symptoms above? Fortunately not, as supplementation with the right forms of zinc can largely 'correct' these symptoms returning you to more optimal body and mental function. For normal chronic zinc deficiency caused by too little zinc in your diet, such as a junk food diet, or less severe genetic imbalances you need only a good quality zinc supplement containing a number of forms of zinc to offset this deficiency and eliminate these symptoms, which is discussed below. Following a period of zinc supplementation and/or change to a zinc rich diet, you can usually discontinue supplementation, but be aware this may take from several months to a half a year or more, depending upon how depleted your reservoir pools of zinc have become (See Fig. 5).

In contrast, if you suffer from kryptopyrroluria, or think you might suffer from this condition, you can take a urine test that can confirm whether you have this condition or not. If you do, fortunately there is a special nutriceutical supplement containing a large number of specific components in addition to high levels of zinc and vitamin B_6. If diagnosed with kryptopyrroluria, once you begin taking this nutriceutical usually for only a period of weeks to a month or so, most or at least many of the symptoms associated with kryptopyrroluria will subside or disappear (Table 4). At first you will need a higher dosage, but as your reservoir pools fill you can then generally take a reduced level of this nutriceutical and remain symptom free.

EFFECTS OF ZINC DEFICIENCY: BEHAVIOUR AND FUNCTION

Whatever the source of zinc deficiency, there are observable behavioural and functional consequences. From the research on animals, especially adolescent monkeys, and more recently on children, the most common behavioural symptoms of zinc deficiency are lethargy, less focused attention (hence poor concentration), regular mood swings and difficulty separating from the mother. Zinc deficiency has also been shown to be involved in ADHD, with a recent double blind study showing that ADHD children improved after receiving zinc supplementation as an adjunct to methylphenidate (Ritalin). And this was using zinc sulfate, one of the least effective forms of zinc supplementation for the brain.

The functional symptoms of chronic zinc deficiency include learning and memory problems, especially short-term memory problems including working memory and behavioural problems, most notably with concentration and impulsivity. This is not surprising as levels of zinc are highest in the frontal lobes and gray matter of the cerebral cortex, and the hippocampal short-term memory areas. The frontal lobes, site of your executive functions such as rational thinking, analysis and understanding cause and effect, allow you to delay gratification and control your social behaviour to stay within the bounds of social norms by inhibiting more impulsive reactions to various stimuli.

The memory problems are often expressed as an inability to retain previously learned tasks and there are often difficulties learning new things. Many people who are chronically zinc deficient, especially if there is a family history of chronic zinc deficiency, have a particular type of memory problem – they can remember numbers, but *just cannot remember names*. A friend to whom I told this after she tested chronically zinc deficient said – 'Boy is that true! I've never been able to remember names! For example, not two hours after my wedding when I went to introduce my husband to my aunt,

I said 'Auntie May I'd like you to meet my husband – what's your name?'

I have worked for years with a German medical doctor, Dr. Gerhard Otto, whose clinic specializes in the treatment of children with learning difficulties. He sends me the children that did not respond to his treatment, the Sunflower® Program. While the Sunflower® Program works for well over 50% of these children, there remains a group who do not respond to this treatment. I have developed an acupressure based program, the Learning Enhancement Acupressure Program or LEAP®, that can generally correct most difficult specific learning difficulties, and building upon the solid structural base provided by the Sunflower® Program, I can usually significantly improve the learning abilities of this recalcitrant group with the application of the LEAP® treatment.

However, there remains a subgroup of children who do not respond to the combined Sunflower®-LEAP® treatment or who only partly respond to this treatment with many improving in some areas, but remaining ADHD with persistent concentration problems. 100% of these children are chronically and often even severely zinc deficient both upon the basis of the zinc taste tests, and zinc blood levels. While it could be purely coincidence, I feel there is a strong correlation between this chronic zinc deficiency and their persistent learning problems. Since their problems are exactly the same problems seen in children with *absolute* zinc deficiency, poor concentration, poor short-term memory, working memory problems often resulting in dyscalculi and mood swings, it seems a reasonable hypothesis.

A Key to ADHD Correction For Some People
Functional Symptoms of Zinc Deficiency

The concentration problems and impulsivity go together as a complex and often present or are diagnosed as severe concentration problems and/or Attention Deficit Hyperactivity Disorder (ADHD). While certainly not all ADHD is a result of kryptopyrroluria, there appears to be a subset of children and adults for whom this is indeed the case. Indeed, for 85 to 90% of the combined Sunflower®-LEAP® treatment resolved the learning and behaviour problems of the children we have treated. Many of the children that were taking Ritalin for concentration problems and ADHD symptoms before this treatment could stop taking medication and maintain normal levels of concentration and activity.

In an attempt to resolve the concentration and ADHD symptoms presented by this subgroup not responding to the combined Sunflower®-LEAP® treatment, we had a number of these children tested for kryptopyrrole. Most the children testing positive for kryptopyrrole also still tested deficient for zinc, even after months to up to a year of high levels of supplementation. After a month or two of supplementation with a special nutriceutical specifically designed to address the nutrient deficiencies created by kryptopyrroluria, all of these children responded positively, with concentration greatly improving and ADHD symptoms decreasing significantly, and in a number of cases totally disappearing.

The other functional problems of chronic zinc deficiency are insomnia and related sleep disorders and the eating disorders anorexia and bulimia. A number of studies beginning in the 1980's linked these disorders to chronic zinc deficiency and later in the 1990's confirmed the link.

Studies then showed that zinc supplements decreased depression and anxiety in anorexia and produced weight gain. Bulimia also demonstrated chronic zinc deficiency as well as vitamin B_6 and tryptophan deficiencies and again supplement with all three of these nutrients appears to lead to improvement. A word of caution, both anorexia and bulimia are complex conditions with strong psychological components, so do not think they are *only* a simple nutrient deficiency.

THE ROLE OF SUPPLEMENTATION IN MAINTAINING ADEQUATE NUTRIENT LEVELS

If you suffer from micro-nutrient deficiency like the zinc deficiency discussed above, there are several approaches to resolving this problem, depending upon the cause of the problem. If you simply eat too few zinc-rich foods and/or foods containing high levels of phytates inhibiting zinc absorption such as fast foods, you can just substitute more zinc rich foods into your diet and perhaps cut out of your diet some of the foods containing zinc inhibitors like processed snacks and junk food.

However, for most people, taking zinc supplements is probably the most efficient and effective way to overcome the consequences of zinc efficiency! This is especially true for those of us with genetically less efficient zinc transporters, for whom getting sufficient zinc from our diet alone is difficult.

Whether you have too many phytates in your food (common in many cereal based fast foods), just don't eat enough zinc-rich foods like red meat, or your genes don't allow you to absorb and utilize zinc well, then by increasing the amount of zinc in your diet through supplementation you will enable your body to take up more zinc and reduce or even eliminate this deficiency.

ALL ZINCS ARE NOT THE SAME: WHAT TYPE OF ZINC SUPPLEMENT?

'Yes, I should supplement with zinc, but which type of zinc?' Zinc supplements like supplements for many other micro- and macro-nutrients come in many forms. Currently most people are not aware of three important points:

1. The form or type of supplement determines its effectiveness, and effectiveness varies widely between different forms of the same nutrient (e.g. zinc sulfate versus zinc gluconate);
2. Different parts of the body require different types of the same nutrient (e.g. liver function appears to require zinc arginate, zinc oratate and zinc citrate in preference to zinc sulfate or zinc gluconate, while the brain appears to prefer zinc gluconate to other types of zinc);
3. Supplementation with only a single micronutrient like zinc, even if is the correct type, may not be effective without a matrix of supporting nutrients (e.g. vitamins A, B_6, and E and other minerals such as iron, magnesium, calcium, etc.)

Again while this discussion is directed toward zinc, the same factors apply to many other nutrients in the body.

Zinc may be attached to many different types of molecules, and the molecule it is attached to determines the dynamics of the zinc ion in the body, whether it will be taken up by a specific tissue or not, and how effectively it can be utilised within the cells. Remember the different types of zinc transporters coded for by Allele 1 and Allele 2, etc.?

One of the most common forms of zinc supplements is *zinc sulphate* as it is cheap to make. Although it is rapidly taken up into the blood, it is poorly utilised by many types of cells in the body, especially the brain. A recent pilot study found that when zinc-deficient children with learning problems were supplemented with zinc sulphate they did indeed increase their blood levels of zinc, but the intracellular zinc concentrations remained largely unchanged. In contrast, while zinc gluconate may be more rapidly assimilated into brain cells supporting mental function and memory, but, as with all forms of zinc, it first needs to get into the blood via a specific zinc transporter. It is this step that may be inefficient for many people thus requiring relatively high concentrations of zinc gluconate to produce even normal levels in the brain cells.

The amino acid chelate, *zinc arginate*, while not used efficiently by the brain, strongly supports the immune system, liver function and digestive enzyme function. For the production of cellular energy, *zinc oratate* can be rapidly taken into the mitochondria from the intracellular fluid by the Zn-T6 transporters, and *zinc citrate* appears to facilitate digestive and liver enzyme function. Other forms such as *zinc picolinate*, are common in many zinc supplements and said to enhance zinc uptake, yet picolinic acid is not found in appreciable concentrations in nature, and it is unknown at this time how well they are transported. One study actually suggested that zinc picolinate negatively affected the zinc status in rats.

Several studies have shown that zinc supplementation alone is often ineffective with no differences observed between placebo and supplemented groups. However, studies that have used zinc supplementation plus a matrix of supporting nutrients have generally shown significant changes in children's behaviour with regard to activity, memory and attention.

Clearly, to truly 'balance' your zinc needs when you are chronically zinc deficient, you must have a broad based supplement containing many different types of zinc. Unfortunately, these supplements are few and far between, as most that are commercially available contain only one type of zinc, and often the cheapest type, *zinc sulphate*. The same is true for many other nutrients like calcium, where the most common form in many supplements is calcium carbonate – cheap to make, but very poorly absorbed.

Unfortunately there are few zinc supplements on the market that have more than one form of zinc, while several forms are required to address the needs in different tissues. However, there is now a nutriceutical developed by direct biofeedback that addressess of all of these concerns. A nutriceutical is a complex nutritional supplement designed to address a specific nutritional problem – in this case zinc deficiency. It is called *ThinkingAdvantage Organic Zinc* and consists of a number of different forms of zinc in an organic matrix, plus a complex of ancillary vitamins and minerals developed specifically to address all of the problems created by zinc deficiency – all in one supplement.

Supplementing with *Organic Zinc* may rapidly restore

zinc levels to normal, addressing all of the imbalances in the immune, digestive and liver detoxification system while providing optimum support for memory and higher-level thinking, your executive functions. If you suffer from kryptopyrroluria, however, you will need a nutriceutical designed specifically to treat this genetic condition.

How much is Enough to Optimize Mental Function?

From the discussion in Chapter 2 it is clear that optimum function will generally not be achieved by consuming only the RDA of zinc or other nutrients, especially when you are already deficient. But why is this so?

In order to explain this further I will start with the basics of nutrient dynamics – there are two types of nutrient pools: **Labile Pools** and **Reservoir Pools**. The labile pool is the pool of nutrients that come in everyday in our diets, and which leave our bodies through excretion, defecation and urination. It is this pool that is supported by the RDAs or DVs. However, what happens on a day when you have not taken in your RDA? The body simply draws upon the reservoir pool of these nutrients in your tissues. But what is a reservoir pool and how does it work? (See Fig. 5.)

When you take in more zinc, or any nutrient, than you need that day, the higher levels in your blood are transferred to the surrounding tissues until the concentrations in the tissues equal those in the blood. So over time, excesses of a nutrient in our food 'charge' our tissue reservoirs so these can be drawn upon when these nutrients are lacking in our diet. Because having constant zinc levels in the blood are

Layers of Cells in Tissue = Tissue Reservoir Pool

Zn	Zn	Zn	Zn	Zn	Zn	Zn	Zn	Zn	Zn	Zn
Zn	Zn	Zn	Zn	Zn	Zn	Zn	Zn	Zn	Zn	Zn
Zn	Zn	Zn	Zn	Zn	Zn	Zn	Zn	Zn	Zn	Zn

Blood vessel = Labile Pool

Zn	Zn	Zn	Zn	Zn	Zn	Zn	Zn	Zn	Zn	Zn
Zn	Zn	Zn	Zn	Zn	Zn	Zn	Zn	Zn	Zn	Zn
Zn	Zn	Zn	Zn	Zn	Zn	Zn	Zn	Zn	Zn	Zn
Zn	Zn	Zn	Zn	Zn	Zn	Zn	Zn	Zn	Zn	Zn

A. Zinc in blood in equilibrium with tissue pools

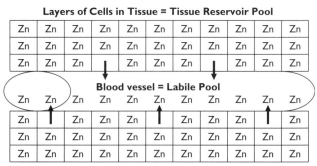

B. Normal – Reservoir pool in tissues maintains blood levels. As blood levels fall, Zn atoms move from tissues to blood.

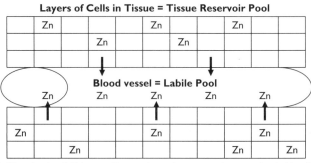

C. Chronic zinc deficiency – Reservoir pools empty trying to maintain blood levels, but levels in blood are still deficient.

Figure 5 Labile and Reservoir Pools of Nutrients. The Labile Pool is found in the blood, while the Reservoir Pool is found in the body's tissues.

important for normal function, blood levels of zinc are tightly controlled by removing zinc from the reservoir pool into the labile pool in the blood whenever blood levels drop. Likewise, when you consume a meal high in zinc, the excess zinc rapidly diffuses into the first level of cells in the tissue reservoirs, recharging your reservoir pool. This is why blood levels of zinc are poor indicators of zinc deficiency, as even very zinc deficient diets do not alter blood levels until the reservoir pools are considerably depleted. (See Fig. 5.)

Once the tissue reservoir pools are exhausted, you are then totally dependent upon your labile pools for maintaining health and normal function. Thus, only when people have chronic zinc deficiency do they begin to show any of the overt symptoms of zinc deficiency listed in Tables 2 & 3 above. People with marginal deficiencies generally have adequate levels of zinc for normal function, but begin to show deficiency symptoms when under stress. Because stress requires extra levels of zinc to meet our needs, we may begin to lose concentration, become forgetful and have difficulty solving problems.

In answer to the question 'How much zinc do I need for optimal mental performance?', you clearly must have more than the RDA of a nutrient to maintain adequate tissue reservoirs for times of stress. The RDA for zinc is currently 15 mgs per day, but levels up to 40 to 60 mgs per day (depending upon your genetics) are probably necessary for anyone wanting to optimize their mental performance because memory and executive functions of the frontal cortex are highly dependent upon adequate levels of zinc.

The labels of many supplements are quite misleading, however, as they often do not give the amount of elemental zinc (the actual amount of zinc atoms), but rather the amount of zinc gluconate, arginate or picolinate, etc. For instance, 60 mgs of zinc gluconate only contains 17 mgs of zinc, the rest being the weight of gluconate. So even taking 180 mgs of zinc gluconate a day is only 51 mgs of actual zinc. Although this is a megadose from the RDA perspective, for people in stressful jobs or in stressful situations in their lives, this megadose may only prevent depletion of their zinc reservoir pools.

Therefore, if you suffer from any of the symptoms discussed above, or just wish to insure that you have adequate zinc levels to maintain optimum function, it may be prudent to consider taking a zinc supplement every day. The best would be a full spectrum supplement that contains all of the different types of zinc needed to support all body and brain functions and a nutrient matrix to synergise the action of the zinc that is taken up such as *ThinkingAdvantage Organic Zinc*.

Key Concept: Types of Nutrient Pools in the body: Labile Pools and Reservoir Pools

The **Labile Pool** is the pool of nutrients that are eliminated from the body each day and replaced by the nutrients you take in your diet. The **Reservoir Pool** is the pool of nutrients that are stored in the cells of your body when there is an excess of a specific nutrient in the blood. The Reservoir Pool recharges the nutrients in the blood to maintain normal function when your intake has not met the demand for these nutrients.

Zinc Toxicity: Myth and Truth

While many doctors incorrectly tell their patients that taking much more than the RDA of zinc is toxic, this simply is not true. The Upper Limit for normal supplementation for a person who is not overtly zinc deficient is 40 to 60 mgs per day of elemental zinc as determined by a recent evaluation of all of the scientific literature on zinc by the Institute of Medicine in the United States. This is more than three to four times the current RDA of 15 mgs per day for zinc.

This 2003 report stated that the Upper Limit is not meant to apply to people with chronic zinc deficiency, which may need a therapeutic dose considerably larger than the Upper Limit to correct the deficiency. Furthermore, the report overtly stated – 'The risks of adverse effects resulting from excess zinc intake from food and supplements appears to be low even at the highest intakes observed.'

Furthermore, the types of reactions resulting from intake of high levels of supplemental zinc, while sometimes uncomfortable, are hardly ever life-threatening. Gastro-intestinal distress (stomach ache) has been reported at doses of 50 to 150 mg per day of elemental zinc, but this is a transitory symptom with no long-term effects. At chronic long-term intakes of 150 mg of supplemental zinc per day, not including the zinc already in the diet, there are some indications of lowered high-density lipoproteins (HDLs), mild suppression of the immune system (a symptom also of zinc deficiency) and reduced copper levels.

However, 150 mg per day of elemental zinc equates to 533 mg per day of zinc gluconate, truly a megadose. Contrast these mild symptoms of zinc '*toxicity*' from *real* megadoses of zinc to the anaphylactic shock that kills several hundred Americans every year from taking a couple of aspirin. Yet many of the same doctors telling their patients zinc is overtly 'toxic' at levels above the RDA are telling them to take aspirin every day!

Chapter 5

Introduction to Brain Integration

Introduction

Most of us will know from personal experience that we can be left exhausted after a period of mentally challenging work and that to some extent 'stress' can clearly interfere with our mental function, memory and thinking. How often have you heard of a child entering an exam room then being unable to perform well on the exam because they suddenly became stressed and could not remember what they had studied and knew at home? Stress and challenging mental work clearly affect our mental performance, partly by burning up the nutrients that feed our brain and maintain its function, but how does it do so?

To understand this, we must first have a basic understanding of how the brain works, what 'stress' is, and thus how it might interfere with our mental performance. The role that nutrition can play in changing our stress levels and optimizing our mental performance is explained in the chapters that follow.

Mental Performance and Brain Integration

Mental performance relies totally upon maintaining integrated brain function under stress – that is, Brain Integration.

Brain Integration is a new understanding of brain function derived from the latest research of how the brain works. In the old view of the brain, different types of thinking and memory were believed to be performed in specific areas of the cortex based on sensory input. Therefore you either accessed these functions and could easily think in certain ways and remember well, or you couldn't.

The new view of the brain is far more dynamic. Thinking and memory are no longer seen as being based in single hierarchical systems, with specific functions performed entirely in one location, but rather are now seen to be widely distributed systems with processing done at many different locations and levels throughout the brain. To provide the processing capacity required of the human brain meant it had to be designed around the principles of Multiplexing and parallel processing.

While multiplexing and parallel processing are highly efficient and provide enormous processing capacity in the relatively small space of the human skull, they suffer one major drawback. Because no function is done entirely in one specific brain region, but rather widely distributed between different brain regions, brain function is 'timebound'. That is, it is dependent upon the synchronization and precise timing of neural flows both within processing centers and along the integrative pathways between these centers to maintain efficient, effective function.

Because even simple mental processes are performed in many different parts of the brain (often at different speeds), the creation of coherent output in the form of thinking requires the integration of all these separate processes. Thinking at higher levels requires even more brain regions to become involved, relying on even higher levels of integration. Thinking can therefore be performed on several levels from simple tasks like 'What is 1 + 1 = ?' to astrophysics and quantum theory.

> **Key Concept: Brain Integration: Underlies all Effective Thinking and Memory**
>
> Thinking and Memory are both highly distributed functions in the brain with the same brain modules involved in processing many different types of information. Effective brain function is thus highly dependent upon the precise timing and synchronization of neural flows within the brain. Loss of synchrony and timing or loss of Brain Integration = loss of some specific mental function or ability. The primary factor desynchronizing neural flows is 'Stress' resulting from activation of the Fight or Flight Reaction of the Brainstem Survival System.

EXECUTIVE FUNCTIONS OF THE BRAIN: THINKING AND ANALYZING

The highest level of thinking is found in the executive, decision-making functions of the frontal lobes of the brain, and thus requires the highest levels of integration to work effectively. An example of a lower level of thinking would be doing simple arithmetic or 'thoughts' about a purely sensory dominated event such as a colourful sunset that may only evoke our emotions, whereas a higher level of reasoning involving say, problem analysis, would cause the 'executive functions' to become involved. These 'executive functions' are housed in the frontal lobes of the brain and rely strongly on our Working Memory in an area of the brain called the dorsolateral frontal cortices situated above and behind the eyes.

It is these executive functions that most clearly separate man from all other animals, since they allow us to take charge over our more basic survival oriented emotions and give us the choice to act or respond in a reasoned way rather than by knee-jerk emotional reaction. Indeed, Elkonhon Goldberg titled his recent book on frontal lobe function, '*The Executive Brain. Frontal Lobes and the Civilized Mind*' to denote the importance of these executive functions in expressing our behavior in socially responsible ways.

The major executive functions of the frontal lobes have been identified as: higher-level reasoning, analytical thinking, multi-tasking, decision-making and problem solving as well as lateral, creative thinking. Higher-level reasoning requires access to the ability to multi-task. This is because to analyze possible outcomes of your actions require the ability to consider them all at the same time. Decision-making is just the ability to choose between the available options. It is also through our executive functions that we can appreciate cause and effect, and thus anticipate the possible outcomes of our actions.

While emotions are necessary to give relevance and meaning to our thinking, we need to control our emotions if we are to think clearly. This control that allows us to make the best possible decisions, even in emotionally charged circumstances, originates in these executive functions. When we lose this executive control, our decisions no longer come from a creative solution-oriented perspective, but from emotional reaction based on either Ego or physical survival. Decisions are then made from the perspective of what is perceived to be good for the survival of the individual, rather than the ultimate good for everyone affected by these decisions in the longer-term.

Our executive functions are the source of our lateral thinking and creative problem solving. These are critical resources needed by every business executive and decision-maker, and if these resources are lost through lack of brain integration, the only option left is emotionally driven, reaction-oriented thinking. When under sufficient stress, solution-oriented thinking is suddenly replaced by survival-oriented thinking focused totally on the now moment, making it difficult or impossible to see alternate choices, or consider the long-term future outcomes of our actions and decisions.

From the perspective of brain function, this change in thinking is reflected as a loss of integrated neural flows, particularly in the frontal lobes supporting rational thinking and analysis, and 'turning on' of our subconscious survival centers for survival.

> **Key Concept: Working Memory, the site of our Executive Functions**
>
> Working Memory located in the Dorsolateral Frontal Cortex (above and behind the eyes) is the site of our Executive Functions such as our Inductive and Deductive Reasoning, Analysis, understanding Cause and Effect, Problem-Solving and Creative and Lateral Thinking. These functions also control our more basic drives and emotions allowing us to follow the Norms of Society and behave appropriately in Social Situations.

SURVIVAL AND BRAIN FUNCTION

Having created thinking and reasoning to solve our problems, for what reason would evolution ever have created a neural mechanism to 'shut-down' our thinking, and leave us at the mercy of our reactive subconscious survival emotions such as fear and anger? And how does this happen?

Survive First, Think later & the Evolutionary Imperative

The first imperative of any animal is to survive so you can reproduce your species. To this end a finely tuned Survival System was built into the brainstem of all vertebrates including man. While our cortex has undergone massive development with the evolution of the Neocortex (or new cortex), and especially the expansion of the frontal cortex in man that houses our executive functions, the Survival System of the Brainstem and the Cerebellum has remained relatively unchanged – protecting you from harm as it did your ancestors (indeed the very reason you are here reading this today).

The Survival System is centered around the core of our brainstem and 'in-between-brain' or diencephalon – that is, in between the brainstem and the cortex. The components have such exotic names as the Periventricular Gray Matter and Periaqueductal Gray Substance, the PVG and PAG respectively, and the Periventricular Hypothalamic Gray Matter or PHG. The first two are the survival centres that run our behavioural reactions for survival, the famous Fight or Flight Reaction, and the PHG is the subconscious centre that elicits the physiological reactions that accompany these behaviours. The last piece of the Survival System is an almond-shaped nucleus about the size of the last joint of your little finger in the medial temporal lobe, called the Amygdala (means almond-shaped in Latin).

The Amygdala is the sentinel of the Survival System constantly surveying all in-coming sensations for potential danger or threat! Indeed, the Amygdala actually makes its own coarse-grained images of objects before the cortex has finished making its more fine-grained, sharp images of the same objects. Upon the basis of this rapid coarse-grained image the Amygdala asks only one question – 'Could the object I see be potentially dangerous?' If it decides the answer is 'No!', the object is ignored, but if the answer is 'Yes!', then it activates the brainstem components of the Survival System to do their thing!

PVG and PAG fire to activate the Fight or Flight Reaction, and you either run to escape, or turn to fight, while at the same time the PHG fires releasing adrenalin and fires the sympathetic nervous system that opens your eyes wide, fully dilating your pupils – to see all possible escape routes, increases blood pressure and heart rate, and further releases more adrenalin to increase the speed and power of muscle contractions for the running or fighting that may follow. Also your mouth may go dry and you experience the survival emotion of 'Fear'. While all of these initial responses are subconscious and have already happened by the time the Fear arrives on the scene, it is the Fear that notifies your consciousness – 'There's danger or threat lurking – Watch out!'

So when you are suddenly confronted with a potentially 'dangerous' stimulus, e.g. you are enjoying a nice walk in the woods when you step forward and there is a 'rattling-rustling' sound and a twisted object in the corner of your eye suddenly

> **Key Concept: The Fight or Flight Reaction of our Brainstem Survival System**
>
> When confronted with a 'potentially' threatening stimulus or situation, the Amygdala 'fires' the Brainstem Survival System activating the Fight or Flight Reaction. This is a cascade of subconscious physiological reactions originating in the Brainstem releasing the Stress Hormones Cortisol and Adrenalin from the adrenal glands as well as activating the Sympathetic Nervous System providing both the energy and stereotypic behaviours to Fight if you need to or Flight-Escape if you can. Activation of this reaction actively 'inhibits' the frontal lobes because 'turning off' our thinking, because Thinking is just too slow for Survival!

appears to rise up at you – the Amygdala perceives possible danger – 'SNAKE!' This instantly fires PVG and PAG preparing you to fight or flight, so you jump back in fear with your adrenalin pumping. Then from your new relatively 'safe' position you turn your head to look more closely – 'What is it really?' There are two possibilities:
1. It is a snake and a rattlesnake at that, so you continue your retreat and perhaps pick up a stick to protect yourself.
2. It is only a twisted piece of vine and you had stepped on one end and the other end rattled and popped up from the ground.

In the first case, you would continue your Fight or Flight reaction. Fight – attack and kill the snake with your stick, or

Flight – run away, depending upon how brave you are. In the second case, you would now realise there was no 'real' danger – just a 'trigger-happy' Amygdala, and you would tell yourself – 'Oh!, it's only a twisted bit of vine.' – laugh and continue on your way as your adrenalin levels recede and your blood pressure and heart rate return to normal.

What happened was your frontal lobes upon closer, more fine-grained inspection 'knew' the object was not dangerous and turned off your initial Amygdala reaction thus shutting down the Fight or Flight reaction. Indeed, this is how it should work – 'react first' to survive then, once you have survived, apply your thinking and knowledge of the world to see if this survival reaction was really necessary or if it was just jumping to conclusions without enough information. (See Fig. 6.)

Remember, the Amygdala always errs on the side of caution – 'If it *could* be dangerous, the Amygdala treats it as if it *is* dangerous!' Then if further inspection proves it's a false alarm, the frontal lobes can turn off the Fight or Flight reaction, as to continue it would be an unnecessary waste of energy. (See Fig.6. next page.)

Staying Alive and Learning

While the Survival System may keep you alive by rapid Fight or Flight reactive decisions to potentially dangerous stimuli, it does not think, it is only associative, and thus cannot 'learn' in the more rational sense of understanding cause and effect. Normally these Amygdala survival associations are correct – 'poisonous snake – danger – Flight', but occasionally they can be totally spurious – just an inappropriate association of a neutral stimulus with a survival emotion, often 'Fear!'

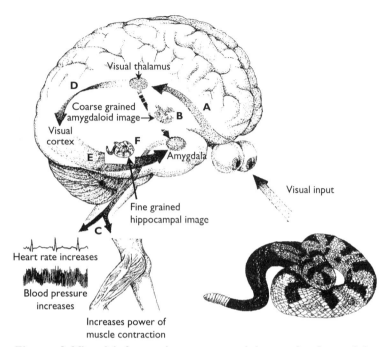

Figure 6 Visual information goes straight to the Amygdala (A) that forms only a coarse-grained image based on rapid processing involving only a few neural links. The Amygdala 'sees' a twisted object on the ground (B) – references 'snake' – danger, sending signals to the brainstem Survival System to initiate the 'fight or flight' response. Signals are then sent to the adrenals (C) to release adrenalin increasing heart rate, blood pressure, and the power of muscle contraction, and you may jump back to avoid the object. At the same time, visual information is also travelling back to the Visual Cortex via the Optic tracts (A) and Optic Radiations (D) from the visual thalamus. Once at the Visual Cortex, this information undergoes multi-step and multi-level processing involving many neural links, to form a fine-grained image of the object (E) that is then referred through the Hippocampus (conscious short-term memory centre) to other long-term memory areas for final identification as a consciousness perception. (Redrawn after J.E. LeDoux, Emotion, memory and the brain. *Scientific American*, June 1994.)

This is the basis of phobias – an irrational association on the part of the Amygdala with some neutral stimulus, but once made, is difficult to change. This is because whenever the stimulus is present, the Amygdala 'sees' it first before you are even conscious of what it is, and pre-emptively fires the Fight or Flight reaction. If this reaction is strong enough it actively 'shuts-down' your frontal lobe rational thinking and all you can do is react in 'Fear', even when at a higher conscious level you 'know' the stimulus is not really dangerous!

The reason higher levels of Fear totally inhibit our thinking is because in life or death situations the people who thought first and acted second didn't leave a lot of offspring – thinking is simply too slow for physical survival! In contrast, the people who 'reacted' first and did whatever kept them alive last time, survived more often. However, once you have survived, then thinking again about what happened (retrospection) and thinking about how you might do it differently in the future (introspection) broaden your survival possibilities and allow you to 'learn' from these survival experiences to provide new options in the future. This has clearly been a successful strategy as witnessed by the number of people in the world today!

Survival of the Two of Us

While all animals share the brainstem survival system for maintaining physical survival (that is staying alive), only humans and probably to some degree the higher primates and cetaceans (whales and dolphins) are self-aware. That is, having a sense of themselves – 'Me' as opposed to 'You'! The

Me that I recognise *is* separate from You. My unique personality – often called my 'Ego'. So with humans there are two things must survive: 1) the physical body, and 2) the Ego-sense of 'myself'. While the Brainstem Survival System takes care of physical survival, it is the newer but still ancient Limbic System that takes care of Ego survival.

The Limbic System evolved to deal with the more complex social interactions of mammals necessitated in part by the suckling of relatively helpless young, and the more cooperative social interactions found in higher mammals. While the physical survival emotions of 'Fear and Anger' are innate and hard-wired into your brainstem, social survival emotions such as 'Guilt and Blame' must be 'learned' as societal mores and norms of social behaviour – software you install as you grow up in a particular society.

So if someone should 'trigger' the survival emotion of 'anger', this subconsciously elicited survival emotion must then be modulated by your Limbic System to be expressed in a socially acceptable way. Depending upon the society, social situation and relative rank of the people involved it may be expressed appropriately in a number of ways. If you were the Queen of England, you might say, '*We* are not amused!' But if you were an Italian, you might say, 'WHAT ARE YOU TALKING ABOUT, I CAN'T BELIEVE YOU SAID THAT!!!' And if I am your friend and 'trigger' your anger, you may yell at me, but I am your boss and you're due for a pay raise, you may well suppress the same anger.

Therefore, part of the development of the frontal lobes and thinking was not just to problem-solve, but also to be able to modulate and control our more basic survival emotions to match appropriate norms of behaviour in social circumstances for social survival, the survival of your Ego. If your boss yells at you and calls you a 'Dumbo!', he is not threatening your life, but he certainly does fire your 'Ego survival' reactions, which once activated, release the same brainstem survival emotion of 'anger' as a 'real' physical threat or danger. If your Ego is attacked you will either react with Fight (displaying 'anger') or Flight (trying to run away if you can) or you will 'submit' and suppress your anger if you cannot.

Thus, an important part of our Ego survival depends upon parts of the Limbic System and especially the frontal lobe executive functions to control and suppress our basic survival emotions. But, if these survival emotions become too strongly activated, we revert to the more primitive expression of overt Fight or Flight and these survival emotions become now fully expressed – perhaps much to our chagrin as we may now say something or do something we wish later we hadn't! However, to suppress strong survival emotions once activated, requires considerably conscious effort and control, and competes with the energy required to problem-solve and think creatively – the reason you become so much less creative in an argument.

Survival and the Loss of Brain Integration

As discussed in the beginning of this chapter, effective brain function is totally dependent upon the synchronised timing of neural flows in the brain, especially those flows that link the various processing centres in the brain – the integrative pathways. The largest integrative pathway in the brain is the Corpus callosum, a sheet of neurons running from the front of the brain to the back of the brain connecting the two cerebral hemispheres into one integrated system. Anatomically the Corpus callosum is between 200 and 800 million fibres connecting the right hemisphere to the left hemisphere and vice versa, and provides the major neural flows integrating the processing of our senses and our thinking. (See Fig. 7 next page.)

Our brain has two types of thinking: Inductive reasoning and Deductive reasoning, each is powerful in its own right, but far more powerful when integrated together. Inductive thinking allows us to reason from wholes or patterns, and hence these functions are often called the Gestalt functions after the German word for 'pattern or form'. Since they operate outside of our normal verbal awareness we speak of the conclusions derived from these functions as our 'intuition' or intuitive thinking. In contrast, Deductive thinking reasons from the pieces that make up the whole, and is linear, sequential and analytical employing the rules of Logic, so these are often called the Logic functions, and because we can consciously follow each step of the reasoning we call it rational thinking.

The Logic functions are located predominately in one hemisphere, the left side for right-handed people, and the Gestalt functions are located on the opposite side, the right side for right-handed people, while the reverse is true for non-right handed people, that is left-handers and ambidextrous people. Integration of the Logic and Gestalt functions are thus highly dependent upon the timing and synchronicity of neural flows across the Corpus callosum making it the major integrative pathway for the cortex. While integration of our basic senses such as seeing and hearing is also dependent upon neural flows across the Corpus callosum, the executive functions of the frontal lobes, our thinking, are even more dependent upon these flows. Indeed, it is the integration of our Logic and Gestalt functions, of our Inductive and Deductive thinking, that underlies our normal problem-solving and creative thinking abilities. (See Fig. 7.)

One of the first functions affected by the stress created by activation of strong survival emotions is the synchronous flow of information across the Corpus callosum. At some level of 'survival stress' the Corpus callosum 'shuts down' and

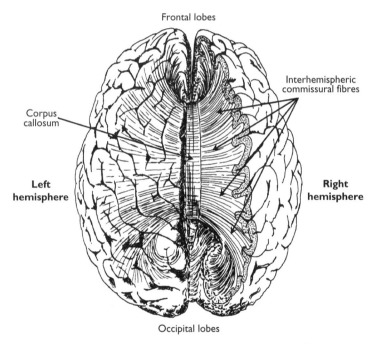

Figure 7 The Corpus callosum – expanded view. On the right side of the diagram the cortex has been removed so that you can see that most of the interhemispheric fibres that cross the Corpus callosum to connect the same area of one hemisphere with exactly the same area in the opposite hemisphere.

the integration of our executive functions is lost, depriving us of our greatest mental resource, full integration of our Logic and Gestalt thinking – the basis of coherent problem-solving. If the survival emotions are even more strongly activated, then the blood flow to our whole frontal lobes is inhibited to actively 'turn off' our thinking altogether to let the faster reactive Fight or Flight system take over for survival. Thinking is just too slow and would actually get in the way of the subconscious reactive survival programs needed to survive.

Shutting down the well-integrated neural flows between the two hemispheres by 'blocking' flow across the Corpus callosum is thus a major reaction to 'stress' and results in the loss of Brain Integration for survival. Unfortunately, the brain does not differentiate between Physical survival and Ego survival, so challenges to our Ego can as effectively 'shut-down' our integrated thinking as challenges to our physical

> **Key Concept: Activation of Survival Emotions 'Shuts Down' Brain Integration by 'blocking' the neural flows across the Corpus Callosum**
>
> The Corpus callosum is the largest integrative pathway in the brain connecting the two cerebral hemispheres. At a 'threshold' level of intensity, activation of the Survival Emotions such as Fear, Anger, Rage or Panic triggered by the Fight or Flight Reaction actively 'inhibit' neural flows across the Corpus callosum disrupting Brain Integration and blocking your Thinking, Reasoning and Rational Problem-solving.

survival. Therefore, a number of factors both physical and psycho-emotional may act as 'stressors' that can 'block' effective thinking and cause a loss of Brain Integration.

Now let us look at the 'stressors' that may 'shut down' our Brain Integration blocking access to our executive functions and leaving our emotional reactions in charge.

Chapter 6

EFFECTS OF STRESS ON MENTAL PERFORMANCE

INTRODUCTION

From the discussion in Chapter 5 it is clear that effective Thinking relies upon coherent well-synchronised neural flows between the two hemispheres, particularly the executive functions of our frontal lobes to link our intuitive and rational thinking for maximum problem solving capacity. Indeed, problems are often solved by a sudden 'flash' of intuitive insight, but only after we have used our rational reasoning to understand what options are available to us. So while the Logic functions provide us with the ability to analyze and understand our problems, it is often our intuitive Gestalt faculties underlying creative and lateral thinking that provide the final solution.

Thus, problem-solving, creative thinking, and sound decision-making are highly dependent upon Brain Integration provided in large part by well-synchronised neural flows across the Corpus callosum. However, this integration is strongly affected by the 'balance' of our survival emotions. When the activation of our survival emotions exceeds a specific threshold, they 'shut-down' or greatly interfere with the flow of information across the Corpus callosum causing a loss of Brain Integration. This loss of Brain Integration then deprives us of our ability to make reasonable, rational decisions and we often go instead into 'knee-jerk' reactive decision-making based on the strong survival emotions being experienced at the time.

At a later date when we are no longer experiencing these strong survival emotions and have regained Brain Integration once more, we may say to ourselves – 'How did I ever make *that* decision, it was clearly irrational, bordering on stupid!' This is merely the consequence of loosing access to your more rational problem-solving and lateral thinking abilities due to the 'stress' of poorly controlled survival emotions.

Factors causing Loss of Brain Integration and Reduced Mental Performance: Effects of Stress on Thinking

It is clear from the above discussion that any loss of synchrony or timing in the transmission of neural flows, especially across the Corpus callosum, will disrupt the mental processes dependent upon them causing a loss of Brain Integration. But what are the primary factors that can cause loss of brain integration?

- Mental stress – often caused by deadlines to meet, with too little time to meet them. This type of stress often called 'Overwhelm' that results from the loss of the executive functions ability to prioritize what you need to do and what you don't need to do *now*, and what order to do them in.
- Emotional stress – commonly results from interpersonal disputes or differences of opinion that activate strong

survival emotions shutting down brain integration and often even the executive functions due to activation of the Fight or Flight reactions for Ego survival.
- Physical stress – one of the most common forms of physical stress is fatigue resulting from over-work and lack of sleep. This often results in the depletion of nutrients and a subsequent lack of sufficient neurotransmitters to maintain brain integration and control strong survival emotions, further shutting down brain integration.
- Physiological stress – activation of the Fight or Flight emotions (fear, anxiety, anger, etc). This causes the release of adrenalin and cortisol thus hyping up the sympathetic nervous system and actively shutting down the frontal lobes for survival.
- Biochemical stress – one of the most common forms of biochemical stress is marginal nutritional deficiencies leading to inadequate production of energy and important neurotransmitters needed to maintain brain integration.

For most people the first three factors causing loss of brain integration are transitory:
- Mentally – you have a big deadline to meet one week, but the next you're on vacation; or
- Emotionally – you've had a very difficult meeting with a person that really pushes your buttons in the morning, but in the afternoon you have a meeting with one of your best friends; or
- Physically – one day you're exhausted from meeting the big deadline the day before, but the next day you feel great after a good night of sleep.

However, the last two factors, Physiological and Biochemical stress, may be on-going. Even though a person may not be aware of being 'uptight', their subconscious emotions may still be activating Fight or Flight emotions. However, these emotions may just not be activated strongly enough to enter their consciousness. And clearly marginal nutritional deficiencies are totally below my level of conscious awareness. I'm only aware of the result, the stress caused by my lack of access to the problem-solving skills and higher reasoning abilities of my executive functions.

STRESS: WHAT IS IT AND WHERE DOES IT COME FROM?

Most people know when they feel 'stress', but what exactly is this 'stress' being felt? First of all, let's look at the different types of 'stress' that can be experienced.

Types of Stress: Psychological, Mental, Emotional, Physical and Physiological

Generally, as we move from psychological to physiological stress, the cause of the stress becomes progressively more subconscious. That is, its origin is less obvious to our consciousness. Normally we are consciously aware of our psychological problems – e.g. we are in a difficult relationship and there appears to be no obvious 'way out' or no easy solution to this emotionally charged problem. This may prompt someone to say to a friend – 'This relationship problem really stresses me out!'

Likewise, when you are working on a difficult strategic plan determining your next year's schedule, you may be aware that this work is mentally stressful. And, of course, physical stress, such as pulling a muscle or twisting your ankle, is usually very conscious because it is painful, and nothing like a little pain to get your attention. In contrast, emotional stress is often subconscious. So with emotional stress, we often feel 'stressed', but are not consciously aware of 'why' we are stressed.

This is because emotional stress originates in the limbic and brainstem survival centres and thus operates much of the time outside of our consciousness. Indeed, a good deal of our psychological, mental and physical stress also has its roots in our emotional survival systems. So while you ascribe your psychological stress to the relationship breakdown, much of this stress originates in a survival system deep in your brainstem, a system called the Panic System that is activated by loss of connection to people. It is here that feelings of sadness, grief and depression are generated, and for that reason, it is also called the Separation Distress System.

Likewise, the stress you consciously ascribe to the mental effort required to sort-out all the complex, and often conflicting events, in next year's schedule probably has a significant emotional component as well. For example, you have this important meeting to close a big financial deal, but it is on the same day as your daughter's birthday, so the feeling of not wanting to disappoint her and missing her big day make the mental act of making this decision stressful.

And although we are very aware of our pain and physical stress, its origins are generally in these same subconscious emotional and survival systems! 'Oh, my neck is so tight and painful I can hardly turn my head!', just happens to occur after working intensely for hours on your next year's schedule, or after dealing with someone who is a real 'pain in the neck!'

The Physiology of Stress: The Reason you know you feel Stressed!

Physiological stress is actually the origin of most of the symptoms we experience as our mental and physical stress. However, physiological stress is biochemical in nature and a direct result of our emotional states, particularly activation of our survival emotions, and thus subconscious in origin.

When the survival centres in the brainstem and Limbic system fire in reaction to perceived 'threat or danger', they initiate a whole cascade of physiological events that prepare the body for 'Fight or Flight!' For humans, this 'threat or danger' can be to either our physical *or* psycho-emotional selves, unlike animals in which this system is only activated to stay alive.

A satiated lioness, after the stress involved in chasing and wrestling her prey to the ground, then defending her share from the other lions while eating her fill, does not become mentally and emotionally stressed by wondering what the other lionesses think of her bulging belly?' But a well-fed sunbathing woman may very well be 'stressed' by what other women think of her figure! Why?

The answer is that what other members of her social group think about her determines her status in that society, and hence her likely share of the resources available, and more importantly, the resources available to rear her children. While she does not consciously think about this, it is what creates concern over what others think of her!

This is how it all works. When a potentially threatening or dangerous stimulus is perceived by the Amygdala (the emotional survival systems control centre) it activates the Fight or Flight system. This, in turn, activates the following series of physiological events:

1. The Central Nucleus of the Amygdala fires signals to the Hypothalamus causing both hormonal and Autonomic Nervous System (ANS) activation.
2. The ANS stimulates immediate release of adrenalin and cortisol from the adrenal glands, plus a redistribution of blood through out the body.
3. The release of adrenalin increases the heart rate, blood pressure, and release of glucose into the blood.
4. This, in turn, increases the tone or tension of the muscles and increases the power of muscle contraction. (This is why a woman may be able to lift a car off her pinned child!)
5. Cortisol release inhibits the immune system to prevent inflammation, and causes rapid release of glucose needed for energy. The adrenalin and cortisol released also block the 'now time' memory of the hippocampus in favour of reactive memory of what you did last time to survive. In fact, chronic high levels of cortisol from prolonged traumatic stress may actually damage the hippocampus, your short-term memory centre.
6. Blood is redistributed from the digestive system to the heart, lungs and muscles so you can Fight or Flee. In such situations, it makes little difference if you digest your meal if it is your last one, because if you survive you can digest it later! This is why an argument over a meal can give you indigestion, as the

argument activates the Fight or Flight reaction shutting down your digestion.
7. The blood is also redistributed in the brain, turning off blood flow to the thinking areas of the frontal lobes and redirecting this blood to the survival centres of the limbic system and brainstem. This is the reason you totally lose your executive functions and analytical reasoning as soon as you go into Fight or Flight. It is the physiological basis for frontal lobe Executive Function 'shut down'.
8. At the same time, the autonomic nervous system causes muscle tone to increase, especially in the postural muscles of the spine and neck. (Can you remember having a pain in the neck or the achy back after a particularly stressful meeting or interpersonal conflict?)

Indeed, the only way the subconscious emotional system has of alerting our conscious mental and psychological systems that there is something to attend to *is* through activation of the physiology of stress. However, chronic stress may dip below our mental horizon and we cease to even be consciously aware of 'how' stressed we are. You can often see by the muscle tension and facial expression that someone 'appears' stressed to you, but when asked if they are under stress, they may well say 'No, why do you think so? I'm doing fine!'

Particularly if these subconscious emotions have been going on for some time due to suppressed survival emotions, the body compensates and the person no longer even recognises that they are under stress. It's like the old Woody Guthrie song – 'I've been down so long it's beginning to look like up to me!'

Biochemical Stress: The Role of Marginal Nutritional Deficiencies!

Marginal nutritional deficiencies, especially of the essential amino acids needed to make neurotransmitters and essential fatty acids necessary for the stability and effective function of neuronal membranes, may commonly result in loss of brain integration, and subsequent 'stress' because of the lack of executive functions needed to provide solutions to problems confronted each day.

However, as you will see in the chapters that follow, it is not just amino acids and fatty acids that are required to maintain effective Brain Integration, but also a host of other nutrients that are involved in the conversion of these vital molecules into neurotransmitters and neuronal membranes. To do so requires not only the precursors, essential amino and fatty acids, but a complex matrix of nutrients: various vitamins and minerals, sometimes including trace elements, to support each step in the conversion of the raw nutrient into its finished product.

STRESS: THE TRIGGER FOR THE SWITCH FROM PROBLEM-SOLVING TO KNEE-JERK REACTION

It is clear that stress appears to be the primary trigger for most people to the switch from their Frontal Lobe solution-oriented thinking to their Limbic-Brainstem problem-oriented reactions. This represents the switch from a mentally resource rich state of higher-level reasoning, analytical thinking, multi-tasking and lateral creative thinking into a knee-jerk reactionary state based on what we did last time to survive. This compromises both the quality, and very often even the quantity of our decision-making.

Clearly the greater the loss of our creative problem solving skills, the poorer the quality of our decision-making, and as integration is lost only a limited number of choices are perceived, with many other choices are not even considered. Equally important, the common response to this limited number of choices (none of which looks particularly effective) is to procrastinate in decision-making, often resulting in last minute rushed decisions or the failure to make timely decisions.

> ### Key Concept: Stress Triggers Switch from Problem-solving to Knee-jerk Reaction
>
> **Stress is a primary cause of loss of Brain Integration.** When stress levels reach a critical threshold it fires the Fight or Flight system activating survival emotions that actively inhibit frontal lobe functions, and directly compete with the problem-solving frontal lobes for limited nutrients required to produce neurotransmitter levels needed to maintain brain integration under stress.

Three Problem-Solving Scenarios: How Loss of Brain Integration affects Mental Performance

In a sense we can think of three scenarios relating to how 'Stress' affects our problem-solving and decision-making abilities:
1. Frontal Lobes are 'turned on' and your Executive Functions are fully intact permitting rapid decision-making and effective, creative problem solving. At the same

time, the survival emotional centers controlling Fight or Flight remain balanced. In a sense they are 'on standby', creating little resistance to Frontal Lobe function.
2. Frontal Lobes are 'turned on' and your Executive Functions are fully intact, but the survival Fight or Flight emotional centres are also in a sense 'turned on'. They are now actively engaged in processing survival emotions, like fear, anxiety or anger. This diverts part of your Frontal Lobe resources to regulating and modulating these survival emotions. While this still allows you to operate out of your solution-oriented Frontal Lobes, it may greatly reduce the speed of your analytical thinking and effectiveness of your decision-making.
3. Frontal Lobes are 'turned off' so that you have little access to your Executive Functions and have largely lost your problem-solving skills. At the same time the survival emotions are fully 'turned on' and now are almost completely in control of your thinking. You therefore shift to reactive thinking that is driven by your survival emotions like fear and anger. Seeing only the 'problem' and few options, you now make your decisions from a short-term crisis perspective.

Frontal Lobes 'Turned On', Survival System 'On Standby' Mode

Since the raison d'etrè of our Executive Functions is to provide us with efficient, creative solution-oriented thinking to rapidly resolve problems, the Frontal Lobes 'turned on' mode can be considered the 'full-speed ahead' mental operating mode designed to produce rapid high-quality decision-making. If there is little activation of the survival emotions (that is, Survival 'on standby' mode), there is no Frontal Lobe energy expended in over-coming the resistance of strong survival emotions and no competition for nutrients, and hence nothing to prevent optimum problem-solving and decision-making.

In this state, you are in optimum problem-solving mode where solution-oriented thinking dominates and decisions are made quickly with minimum effort. This then creates a relatively stress-free state of function. An analogy might be your foot on the gas pedal, accelerating rapidly down the road.

Frontal Lobes 'Turned On', Survival System 'Turned On' Mode

In contrast, in the presence of active survival emotions (i.e. the Survival 'turned on' mode) some of the Frontal Lobe resources have to be diverted to modulating these survival emotions. In a sense this creates resistance to Frontal Lobe activity. In this Frontal Lobe 'turned on' and Survival 'turned on' mode, this competition for the Frontal Lobe resources may reduce availability of the Executive Functions needed for effective problem-solving and rapid decision-making.

To continue the car analogy, you now have one foot on the gas pedal and the handbrake pulled on. While you may still retain a good deal of your problem-solving skills, and make reasonable decisions, there is now much more resistance requiring more mental effort, and hence higher levels of stress.

Frontal Lobes 'Turned Off', Survival System 'Turned On' Mode

If the survival emotions become too active (i.e. the Survival 'full on' mode) the Frontal Lobes become totally inhibited by these strong survival emotions, and the brain goes into Frontal Lobe 'shut down', so that thinking does not interfere with your survival reactions. This Survival 'full on' and Frontal Lobe 'shut down' ensures that thinking does not interfere with your survival reactions. As there are no longer any creative, solution-oriented functions available, you now react out of the basic survival emotions of fear, anxiety or anger.

To continue the car analogy further, it is like now having one foot on the gas pedal, while the other foot is jammed on the brake pedal, causing you to grind to a standstill as the engine 'fights' with the brakes for control. While this diverts all of your energy to deal with the perceived 'problem', it is bought at the cost of increased levels of stress, the release of hormones like cortisol and adrenalin and increased sympathetic nervous activity. In addition, there is now considerable conscious effort required to control strong survival emotions. Taking all of these factors together, this constitutes overt physiological, emotional and mental stress.

In Pursuit of 'Full-Speed Ahead' – Optimum Mental Performance

How much of the time do you think you are in 'full-speed ahead' mode? From my experience of over twenty years of clinical practice, only a small percent of people 'live' in this optimal mode of mental processing, and it is not surprising that these people are generally at the very top of their respective professions. For the great majority of us this 'full-speed ahead' mode represents only our 'peak' experiences, not our daily mode of operation.

By far, most of us operate in the Frontal Lobe 'turned on' and Survival 'turned on' mode on a day-to-day basis, and thus we squander much of our Frontal Lobe resources on modulating our survival emotions. When peak stress then comes on line, most of us drop into the Survival 'full on' and Frontal Lobe 'shut down' mode – a state of loss of brain integration in which few problem-solving skills remain and stress may overwhelm us.

But how can we stay in Frontal Lobe 'turned on' and Survival 'balanced' mode on a daily basis? One approach is to 'feed' your brain the proper nutrients to eliminate marginal nutritional deficiencies and provide the excess nutrients required to maintain brain integration, even when highly stressed!

But how can just nutrition alone keep us in optimum mental performance mode? The answers lie in optimal nutrition for the brain.

Chapter 7

HOW NUTRITION CAN OPTIMIZE MENTAL PERFORMANCE

INTRODUCTION

If optimum mental performance depends upon maintaining Brain Integration under stress, what role does nutritional supplementation play in the performance of the brain?

Ultimately, brain integration is actually a biochemical phenomenon, as it is dependent on both neuronal membrane stability and the production of a variety of neurotransmitters (NT's) together with their receptors and transporters. These are the molecules that facilitate and transmit nerve impulses from one neuron to another and carry the signals being integrated.

Brain integration will therefore be disrupted and all dependent functions will be impeded or lost if, for any reason, these transmissions or signals fail or are disrupted. The disruption of neural signals may be caused by insufficient levels of NT's and/or receptors or transporters with the most common factors affecting their availability being the lack of either precursor molecules such as amino acids and fatty acids, or the lack of sufficient levels of ancillary nutrients – that is, the nutrient matrix of vitamins and minerals required for the formation of these important molecules. In particular, fatty acids are required to provide stable membranes to carry out neural signaling. Or, of equal importance, is the lack of sufficient nutrients to maintain the high levels of energy production required to maintain frontal lobe function. Thus, a lack of even one of the nutrients necessary to make these signaling molecules, or provide the energy needed for thinking can overtly compromise brain integration.

However, different areas of the brain and even different brain functions in the same brain area use different NT's, receptors and transporters, and thus are dependent upon different combinations of nutrients to fuel them. Likewise, fatty acid composition of the neuronal membranes varies between different parts of the brain, and therefore the deficiency of a specific nutrient will affect those brain functions dependent upon it, but may not affect other functions. Finally, different parts of the brain have very different energy requirements, with the frontal lobes using up to four times the energy of the brainstem survival system.

Nutritional Deficiency: Effects on Brain Integration

Absolute deficiencies (that is, nutrients almost wholly absent from the diet), while at most a minor problem in the developed countries, are a major problem for children in parts of the developing world. A number of studies have shown that just supplementing malnourished children in India and Guatemala with iron and/or zinc plus a basic mutli-vitamin and mutli-mineral supplement actually increases IQ and cognitive performance on tests of both memory and reasoning. Clearly, absolute nutritional deficiencies result in on-going loss of Brain Integration.

In contrast, while absolute nutritional deficiencies are uncommon in the developed countries, marginal deficiencies are widespread, as discussed in Chapter 2. At normal levels of activity and function, even with marginal nutritional deficiencies, nutrient levels are generally sufficient to support basal brain function. However, when marginal deficiencies are coupled with stress, the brain just does not have enough nutrients to handle these 'peak' demands and therefore loses integration. At some critical level of mental activity or intensity of function, the brain simply runs out of enough 'fuel' to maintain optimal function.

For most of the major neurotransmitters, receptors and transporters there are key essential nutrients that come only from your diet, these being the essential amino and fatty acids. Lack of one or more of these key nutrients unfortunately turns off production of the key signaling molecules just when you need them most, causing loss of brain integration and thus reducing your capacity for effective thinking and memory.

Likewise, critical ancillary nutrients such as iron and zinc are often not present in adequate levels in many people's diets to satisfy 'peak' demands. This can then result in the loss of brain integration and optimal mental performance even with the presence of sufficient essential amino and fatty acids in your diet. What is required therefore is a balance of ancillary nutrients with the correct concentrations and ratios of one nutrient to another to ensure sufficient levels of neurotransmitters for 'peak' demands. This balance of different supporting nutrients is called the ancillary *nutrient matrix*.

> ### Key Concept: Role of the Ancillary Nutrient Matrix
>
> Essential Amino Acids & essential Fatty Acids are precursors needed to make Neurotransmitters and maintain Neuron Membrane stability, thus providing the biochemical basis of Brain Integration. These vital molecules can only be made if there is a complete Ancillary Nutrient Matrix, that is, all of the other nutrients required to support the complex biochemical reactions producing these vital molecules. Deficiency of even just one ancillary nutrient can slow down or block production of these important molecules.

Neurotransmitters: The Genetic Connection

Again, the genes you inherit from your parents and ancestors may be a major factor. For correct levels of NT's, receptors and transporter molecules to be produced in the brain, the genes for these important proteins must produce fast, efficient enzymes able to supply adequate NT levels to maintain brain integration, especially when under stress.

For example, Allele 1 codes for a fast and efficient version of the enzyme that catalyzes the conversion of the amino acid tyrosine into noradrenalin, an important NT modulating brain function, especially in the Fight or Flight system. A genetic mutation produces Allele 2, and this mutated gene codes for a far less efficient and slower version of this enzyme. In this case, people with Allele 1 need far less of the amino acid tyrosine in their diet than people with Allele 2 to produce and maintain normal levels of noradrenalin in their brain. As a consequence, people with Allele 2 may not have sufficient tyrosine, even when eating a balanced diet, to maintain 'normal' levels of noradrenalin, and will tend to lose brain integration under levels of stress that pose no problem for people with Allele 1.

Therefore, people with 'good genes' that produce efficient enzymes may not need additional nutrients to satisfy 'peak' demands, while people with 'not so good genes' producing slow, inefficient enzymes probably will. A lack of sufficient NT may then result in the loss of brain integration blocking access to executive problem-solving functions and memory.

A Biochemistry Primer: Production of Neurotransmitters – How it works

Figure 8 shows the basic plan of biochemical reactions in the body. A *substrate*, a precursor molecule, tyrosine in our example below, reacts with a *functional group* in the presence of a catalyst, the *enzyme,* plus the supporting cast of other

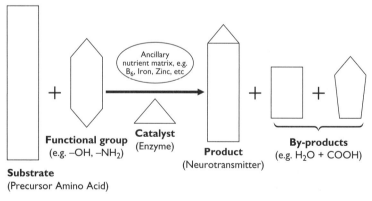

Figure 8 A Basic Biochemical Reaction.

nutrients needed for the reaction to proceed, called the *ancillary nutrient matrix*.

The enzyme catalyzes the joining of the precursor to the functional group producing the *product*, noradrenalin, plus *by-products*. While detailing NT production, the discussion below applies equally well to the production of all receptor, transporter and fatty acid molecules.

Through natural selection, evolution created fast efficient enzymes to produce not only enough NT's and fatty acids for everyday survival, but sufficient reserves to handle times of peak stress. Indeed, life is about the stress of problem solving, and the people who solved life's problems more effectively left more offspring, and these were the genes handed down to the next generation. (See Fig. 9.)

But what do you do if you inherited a gene coding for a slow, inefficient enzyme?

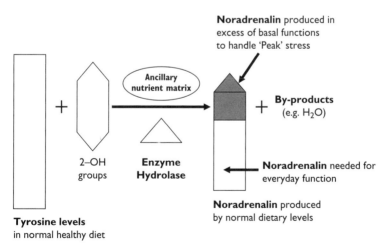

Figure 9 Evolutionary Insurance: Making enough for Stress.

When faced with this problem in the laboratory there are three choices:

1. Choose a more efficient, faster enzyme;
2. Heat the system to make the slow enzyme 'speed up' in order to produce the desired levels of product; or
3. Simply add more precursor molecules to 'drive' the slow enzyme faster.

Unfortunately choices 1) and 2) are not available in biological systems, as you can only work with the genes you have (until biotech produces a faster fish gene to insert into your cells), and our bodies cannot be heated up without dire consequences. So that leaves only choice 3), simply adding enough precursor by supplementation to drive the slow enzyme faster. (See Fig. 10.)

An analogy would be a waterwheel. How fast the wheel turns depends upon the depth of water behind the dam and

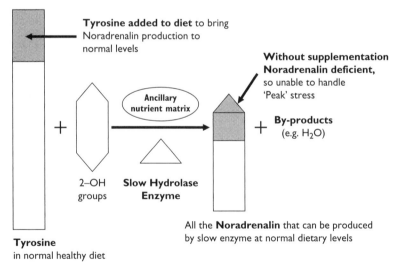

Figure 10 Adding More Precursor: Driving a Slow Enzyme Faster.

> **Key Concept: Evolution created Fast, Efficient Enzymes to provide an Excess of Neurotransmitters for Times of Stress**
>
> Evolution provided not only average levels of neurotransmitters to handle normal brain function, but an excess of these vital signaling molecules to handle times of peak demand, such as times of stress to keep our thinking and problem-solving resources 'online' when we need them the most! When the available levels of neurotransmitters are exceeded, we lose integrated brain function supporting our thinking and problem-solving, and revert to knee-jerk reactive programming from our subconscious survival programs.

how well the wheel has been constructed and maintained. The wheel is analogous to the role of an enzyme, turning the grinding stone to make the product, while the depth of water behind the dam is like the amount of precursor molecules available to drive the enzyme.

Mill 1 can grind the normal 10 bags of grain a day with one meter of water behind the dam because it has a waterwheel that is well aligned and well greased. However, the mill owner builds a dam with not just one meter of water, but three meters so his mill can handle not only the peak demand of 20 bags of grain a day during the 'stress' of harvest, but also so he has a reserve in case of emergencies, and thus is able to continually satisfy all of his customers.

With the same three-metre depth of water, Mill 2 can only grind a maximum of 10 bags per day because it has an old misaligned waterwheel with a rusty bearing. During the 'stress' of harvest time, this mill's owner just cannot keep up with his customers' demands, and has trouble producing enough grain to keep up with his more efficient competitor. However, if he raised his dam to six meters he too could now meet the harvest demand and have excess capacity should something go wrong.

Clearly in this example you would say the owner of Mill 2 should simply realign and grease his waterwheel, and thus not have to pay more to build such a high dam.

When you have a gene coding for slow enzymes, you do not have this option of repairing your waterwheel but you can, however, increase the depth of water behind your dam by supplementing with the nutrients you need in excess of normal dietary levels, and thus drive your slow enzyme fast enough to produce the NT's needed to maintain normal brain function even under stress.

Fatty Acids: The Key to Membrane Stability and Optimal Function

Another critical nutrient for optimal mental performance is fatty acids (FA's), as much of the brain is made up of fatty acids. The white matter of the brain is made up largely of myelin, a whitish fatty acid that insulates axons for high-speed conduction, also why the white matter is white.

All of the membranes of the billions of neurons in the brain are also made of fatty acids, and effective nerve conduction requires the correct type and amount of fatty acids to maintain membrane stability and function. Because Brain Integration relies upon effective, synchronised nerve conduction, it also relies directly upon sufficient concentrations and correct ratios of fatty acids being present in the brain.

While fatty acid biochemistry is very complex, I will summarise the salient points with regard to fatty acids in the brain and the roles they play in optimal mental performance.

Fatty Acids in the Brain: What is the Correct Type?

There two major groups of fatty acids: saturated and unsaturated fatty acids that differ in that the unsaturated fatty acids have one or more double bonds in their structure while saturated fatty acids have none. Because hydrogen atoms saturate all of the bonding sites along the carbon chain of saturated fatty acids, these molecules are relatively 'stiff' and not very flexible as fatty acids go. While this is good for structural integrity, it poses problems for functions like neural signalling where membrane flexibility determines efficiency.

In contrast, unsaturated fatty acids have one or more places where a hydrogen atom is missing and two bonds connect two adjacent carbon atoms forming a 'double bond'. (See Fig. 11 overleaf.) The presence of this relatively more flexible double bond means membranes containing unsaturated fatty acids are more fluid and have different properties than membranes made of saturated fatty acids.

Unsaturated fatty acids come in several flavours depending upon how many double bonds they possess and where these double bonds are located. Monounsaturated fatty acids like olive oil and evening primrose oil have only one double bond, while polyunsaturated fatty acids like the Omega-3 and Omega-6 fatty acids have more than two double bonds.

NUTRITION FOR THE BRAIN

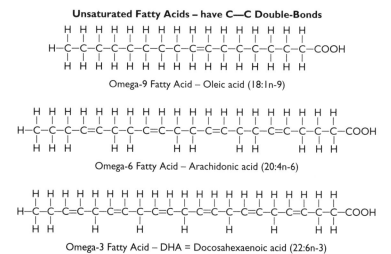

Figure 11 Saturated and Unsaturated Fatty Acids. Note presence of C=C double bond in Unsaturated Fatty Acids. Number in parentheses gives number of Carbon atoms: then number of double bonds, then Omega number (e.g.-9, -6, -3).

Figure 12 shows that Omega-3 and Omega-6 fatty acids differ in the location of their first double bond with the Omega-3 fatty acids having their first double bond at the third carbon atom from the end, and Omega-6 having it

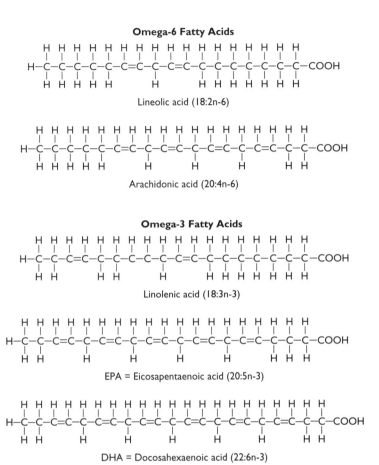

Figure 12 The Omega-3 & Omega-6, Unsaturated Fatty acids.

at the sixth carbon atom from the end. All polyunsaturated fatty acids are flexible, but Omega-3 fatty acids are much more flexible than Omega-6 fatty acids, and Omega-3 fatty acids predominate in the cortex and at the synapses, the junction between neurons.

The most unsaturated fatty acid in the body, Docosahexaenoic acid (or DHA for short) is an Omega-3 fatty acid with 22 carbon atoms and 6 double bonds, and thus is the most flexible fatty acid in the body. The brain is 60% fatty acid with 25% of this being DHA, in contrast to other tissues of the body that are between 0.2% and 4% DHA. But the amount of DHA also varies within the brain with the cortical gray matter being approximately 40% DHA, and the retina of the eye having the highest concentration with greater than 90% of the fatty acids being DHA.

Replacing DHA in the retina with Eicosapentaenoic acid (EPA), another fatty acid having just one fewer double bond, decreases signaling intensity by 50%. Also, a deficiency of DHA in the diet can make the retina 10 times less sensitive to light and decreases the signalling capacity of neurons throughout the brain. Animals experimentally deprived of DHA in their diet grow to be slow-witted and half-blind.

Omega-3 Fatty Acids: EPA and DHA

The Omega-3 fatty acid DHA, along with EPA, play important roles in the brain, while EPA is one of the most important Omega-3 fatty acids in the body, and especially important for the circulatory system. (See Fig. 12.) EPA can be converted into DHA in the brain, but only slowly and inefficiently. Too much EPA relative to DHA can, in fact, decrease neuronal membrane efficiency because when EPA replaces DHA in neuronal membranes that then slows neural signalling.

Thus, the proper ratio between the EPA and DHA is essential for optimum brain function, and the brain, especially frontal lobe functions, requires more DHA than EPA. While having both EPA and DHA in your diet is important, the ratio of EPA to DHA is equally as important, a fact not generally recognised in the nutritional literature.

Sources of EPA and DHA: Ideal and Real

Ideally, the body can convert several other Omega-3 fatty acids like alpha-linolenic acid (ALA) into EPA, and then EPA into DHA. So, technically EPA and DHA are not essential fatty acids. The production of DHA from ALA, however, requires *both* sufficient ancillary nutrient co-factors, that is vitamins and minerals, *and* fast effective enzymes. Clearly, deficiency of any of the ancillary nutrient matrix can slow production and result in DHA deficiency in the brain.

More problematic are the five-enzymatic steps in this conversion, and thus five chances that an Alleles coding for a slow enzyme may disrupt this process. Also, the following factors tend to reduce DHA production by blocking the primary enzymes converting ALA to DHA (the delta-6-desaturase enzymes): high levels of trans fatty acids (common in fast foods), sugar, stress, alcohol, aging and medication like aspirin and other non-steroid anti-inflammatory drugs or NSAIDS.

Both trans fatty acids and excess sugar can interfere with fatty acid synthesis, and the presence of these interfering compounds in our diets has increased 2500% and 250%, respectively, in the past 50 years. For instance, it has been estimated that it may take 100 molecules of ALA to produce 1 molecule of DHA, partly because of all the 'blockers' to proper fatty acid synthesis present in our modern diet. Therefore, EPA and especially DHA production is often so inefficient, that both may be considered conditional-essential fatty acids, that is, fatty acids you must get largely from your diet.

Unfortunately, our Omega-3 fatty acid consumption has also decreased between 80% and 85% in the past 75 years. This is largely a result of the commercial processing of oils that create trans fatty acids, the loss of cereal germ in processed cereals, and the increased consumption of Omega-3-deficient warm weather oils like corn and sunflower oils. (See Fig. 13.)

The best dietary sources of these important Omega-3 fatty acids is cold water fish like salmon, mackerel, herring, sardines and tuna. The ratio of EPA to DHA, however, varies considerably between the different types of fish, and tuna has the highest DHA to EPA ratio making it the preferred source for optimal brain function. The bad news is that many of the fish with the highest amounts of DHA and EPA like tuna and swordfish, are unfortunately contaminated with such

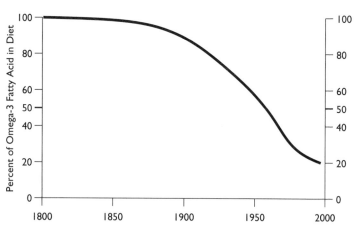

Figure 13 Decrease in Omega-3 Fatty Acids in Western Diets.

high levels of mercury and PCBs (toxic chlorinated hydrocarbons) that the US EPA has suggested people restrict consumption of these fish.

Of course another more accessible and controlled dietary source is supplementation with high quality tuna oil that contains the proper ratio of DHA to EPA. It is important to note that low quality (cheaper) oils may be quite unstable and contain significant amounts of mercury, dioxins, PCBs and other pesticides, and undesirable oxidation products. While salmon oil supplements are far more common than tuna oil, partly because of salmon farming, salmon oil has much more EPA than DHA, and thus is not as desirable a source of Omega-3 fatty acids needed to support optimal brain function. So the best source of Omega-3 oils to support optimum brain function is high quality tuna oil distilled to eliminate mercury and other toxins.

Energy Consumption and Executive Functions

The brain uses a large percentage of all the energy consumed in the body. When you are thinking intensely to figure out a difficult problem, even without the stress of a deadline, your brain may consume more than 50% of all the glucose and 20% of all the oxygen used in the whole body. When you add the stress of deadlines and difficult emotional choices (e.g. I know Fred has five young children, but for the good of the business I have to fire him on the basis of his performance) the brain consumes even higher levels of energy.

However, this energy is not used evenly throughout the brain, but rather is concentrated in the cerebral cortex and, during intense mental activity like decision-making, is largely consumed in the frontal lobes by the executive functions. In contrast, the limbic emotional and survival centers of the brain use less than half of the energy of the cortex in order to function. This is largely because they rely on emotionally generated stereotypical 'hardwired' reactions, not from careful consideration and evaluation of the choices requiring the activation of far more neurons.

To maintain executive control of decision-making during intense mental activity requires sufficient levels of nutrients to maintain high levels of energy production, as well as sufficient levels of precursor and ancillary nutrients to maintain neurotransmitter and fatty acid levels to support these functions. Experiments have shown that when dietary ALA or DHA is restricted, the level of ATPase (the enzyme making the energy molecule ATP) drops by one-half in the nerve endings, enough to slow nerve conduction in the brain.

The critical role of vitamins and minerals in producing the energy molecule ATP is clear when you understand how much less ATP is made from each molecule of glucose when these nutrients are in short supply. Without sufficient amounts of Vitamin B_2, B_3 and the minerals Iron, Copper and Sulfur, only 2 molecules of ATP can be made from each glucose molecule, while when sufficient supplies of these nutrients are available, 34 to 36 molecules can be made from each molecule of glucose. (See Fig. 14 on the following page.)

Figure 14 Nutrients needed to maintain Krebs Cycle Aerobic Respiration in the cell. Note the number of B_2 & B_3 molecules, and the minerals Sulfur, Iron and Copper required to turn one Glucose molecule into 34 to 36 molecules of ATP, the energy currency of the cell.

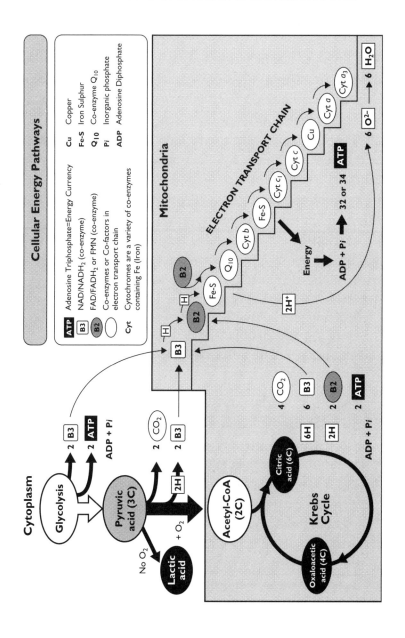

In addition, up to 40 other nutrients are needed in ancillary roles to maintain the enzymes required to run this cellular production of energy. Clearly, deficiency in any of these crucial nutrients necessary to sustain production of ATP may result in an 'energy crisis' in the brain.

When nutrients are in short supply the energy production of the brain is reduced, and the frontal lobe problem solving functions are shutdown to redirect the remaining nutrients to the survival system. So, when push comes to shove and there are limited nutritional resources available, the Limbic-brainstem survival system takes the lion's share to maintain survival, as thinking is both too slow and too demanding of energy to be maintained.

> **Key Concept: Energy Production is essential for Thinking and Problem-solving, and Energy Production is dependent upon the availability of Sufficient Nutrients**
>
> The Krebs Cycle that generates virtually all of the energy for the brain in the form of ATP depends upon two things: oxygen and a complex nutrient matrix of up to 40 different vitamins and minerals to maximize ATP production. Deficiency of any one of these nutrients can slow down energy production, and the Frontal Lobes use more than four times the energy of the brainstem survival system. So when optimum nutrient levels are not available, the Frontal lobes supporting our thinking and problem-solving often 'shut down', and we switch to less energy demanding stereotypic survival reactions to survive.

Chapter 8

NUTRITION FOR OPTIMUM MENTAL PERFORMANCE

INTRODUCTION

Optimal mental performance clearly relies on your ability to maintain Brain Integration in stress situations, and to do so requires many different types of nutrients to sustain neurotransmitter (NT) and fatty acid (FA) synthesis for proper neural signalling, and to convert glucose into the energy molecule ATP to power the brain.

Very often, only those pathways directly involved in feeding the brain are considered, whereas it must be understood that some of these primary pathways are dependent upon other secondary pathways, which are in turn dependent upon yet other pathways. This chain of dependencies means that nutrient deficiency in even an indirect pathway may well slow down or disrupt major pathways that appear to have all of the precursors and other nutrients needed to maintain brain integration.

To put this into technical terms, the biochemical pathways directly involved in NT or FA production rely on precursor molecules or functional groups that often must be produced through other biochemical pathways requiring their own set of precursors and nutrient matrix. Nutrient deficiency any place along this chain may reduce NT or FA levels resulting in loss of Brain Integration negatively affecting your thinking and memory.

Taking Nootropic Herbs, that is herbs that enhance various cognitive and memory functions, may play an important role in optimising mental performance. Herbs like Gingko and Ginseng are well known to many people and have a long history of traditional use to facilitate mental function.

And perhaps the most important factor in maintaining optimum mental function is the *synergy* between the different nutrients supporting and optimizing brain function. That is, the positive re-enforcing interactions between different nutrients, that taken together, work to maintain Brain Integration and create optimal mental performance. This nutritional synergy is commonly overlooked in Western studies that tend to consider or test only one or a few nutrients at a time.

AMINO ACIDS TO OPTIMISE MENTAL PERFORMANCE

To maintain NT levels, especially when under stress there are a number of essential (dietary) amino acids that must be present in abundant supply. The most important amino acids and their related NT's are listed in Table 5 opposite.

Of these amino acid precursors, only the last three, phenylalanine, tyrosine and tryptophan, are essential amino acids that must be provided directly by our diet. On the other hand, Glutamine and Glycine are seldom deficient because these amino acids are abundant in many foods commonly eaten, even in today's Junk Foods. Even though GABA is not

Table 5 The Amino Acid precursors to the major Brain Neurotransmitters

Amino Acid	Neurotransmitter (NT)
L- Glutamine	Glutamate or Glutamic acid
L- Glycine	Glycine
L- Butyric acid	Gamma Amino Buteric acid (GABA)
L- Phenylalanine	Dopamine & Noradrenalin
L- Tyrosine	Noradrenalin & Dopamine
L- Tryptophan	Serotonin (5-Hydroxytryptamine)

found in food, it can easily be made from glutamine in a one-step reaction, and thus these amino acids usually do no pose deficiency problems.

Glutamine is the precursor for *glutamic acid* and *glutamates*, the primary excitatory NT's in the brain that activate various brain areas 'turning them on', while Glycine and *GABA* are the primary inhibitory NT's in the brain, and 'turn off' or 'tune' the brain areas turned on by the glutamate NT's. GABA is the neurotransmitter used by as many as one-third of all the synapses in the brain, and GABA neurons are particularly numerous in the frontal lobes. This may be because it is inhibition that appears to 'tune' the timing of neural signals and is thus more important than excitation in maintaining our brain integration that in turn controls our thinking and behaviour.

Phenylalanine is a primary precursor for both **dopamine** and *noradrenalin* with *dopamine* the primary NT produced from this essential amino acid. Dopamine plays a major role in motivating you to seek reward, but is also important in control of movement. When dopamine is released the Nucleus Accumbens, it makes you 'feel good', a 'reward', and if a lot of dopamine is released by taking alcohol, marijuana, nicotine, heroin or cocaine – you feel very, very good indeed, called getting high!

However, elsewhere in the brain dopamine is a major inhibitory NT modulating and smoothing out muscle action. If the dopamine secreting neurons in the 'Black Substance' or Substania niger begin to die off you get the shaking and tremors of Parkinson's Disease.

Tyrosine is the precursor for both **noradrenalin**, and by an alternate pathway, *dopamine*. *Noradrenalin*, the first cousin of adrenalin is a stimulating NT, one of the major modulatory NT's in the brain. This substance makes us remember exciting or traumatic events by helping activate memory circuits that transfer the memory of these events to long-term memory, and is also involved in sustaining the Fight or Flight reaction. When present at elevated levels, it may lead to 'hyper-vigilance' often resulting in anxiety and sleep disturbances.

Noradrenalin helps to modulate your moods and gives you energy. In contrast, *tryptophan* (the only precursor of *serotonin*) is the major calming NT that to a degree counteracts the actions of noradrenalin. When serotonin levels are too low, depression often results. This is the reason that Selective Serotonin Re-uptake Inhibitors (SSRIs) like Prozac that prolong the action of serotonin are used to treat depression.

However, for you to make adequate levels of dopamine, noradrenalin and serotonin, your diet must contain enough of the essential amino acids phenylalanine, tyrosine and tryptophan. You can easily see how important the precursor amino acids are in the manufacture of these important neurotransmitters from the similarity in structure of each amino acid to the NT from which it is derived. (See Fig. 15.)

As can clearly be seen from the above, these essential amino acids are important for brain function and need to be an essential part of your diet if you are to optimize your mental function. If they are deficient, as they are in many people's diet today, your mental function will certainly suffer to some degree. Thus, at the least it is important to eat foods rich in these amino acids, but to be sure you have enough of these vital nutrients, taking supplements containing these amino acids will help optimise your brain function.

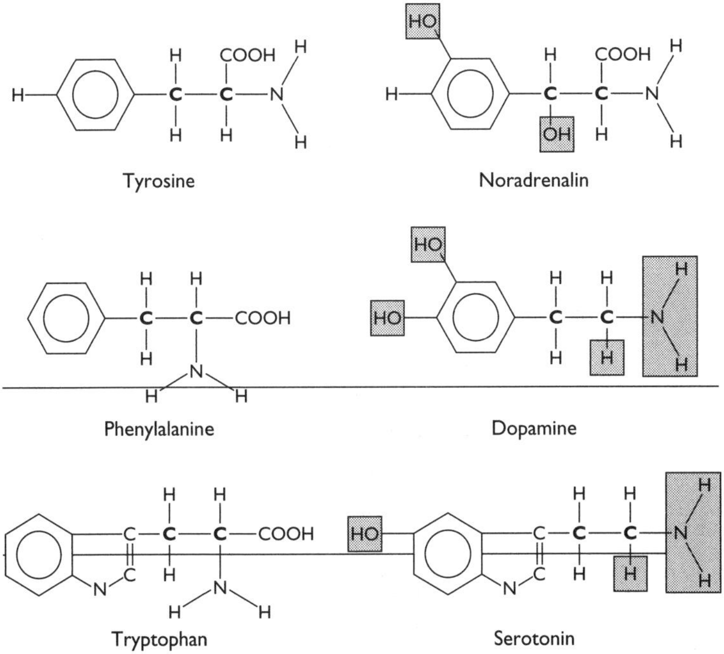

Figure 15 Structure of Essential Amino Acid Precursors and their related Neurotransmitter. Shaded areas differences between precursor and neurotransmitter.

Fatty Acids to Optimise Mental Performance

Saturated fatty acids, such as sunflower and corn oils, and polyunsaturated Omega-6 fatty acids are also important in the structure and function of nerve membranes, but are abundant in most Western diets, in fact often too abundant in many people's diet today.

In contrast, the poly-unsaturated Omega-3 fatty acids are often deficient, especially DHA, the primary fatty acid in the brain controlling neural signaling and receptor activity. However, both Omega-6 and the Omeg-3 poly-unsaturated fatty acids EPA and DHA are needed for effective brain function.

In the past several hundred years, the ratio of Omega-6 to Omega-3 fatty acids has changed dramatically from an original 1:1 ratio of our ancestors. Today the ratio of Omega-6 to Omega-3 fatty acids varies from between 20:1 in general to as high as 45:1 in some mother's milk. (See Fig. 16.) Infant formula commonly has a ratio of 10:1, but until 1997 none of this was in the form of DHA. This changing ratio of fatty acids appears to have serious implication for brain function.

This excess of Omega-6 fatty acids affects Omega-3 fatty acid uptake and metabolism, because competitive inhibition of the Omega-3s by the more abundant Omega-6s reduces the concentrations of Omega-3s available to the cells of the brain. Because of the far lower amounts of the Omega-3s, especially DHA, in most people's diet today it may be important to take a supplement containing these fatty acids to optimise your brain function. Indeed, supplementing with Omega-3 amino acids has been shown to increase the speed

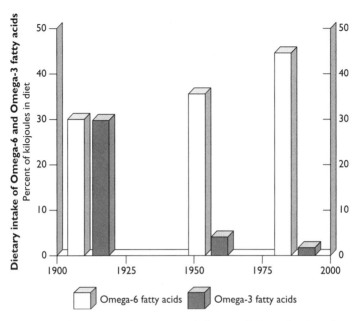

Figure 16 Changes in Omega-6 and Omega-3 Fatty Acid intake in Western Diets over the past 100 Years.

of nerve conduction and visual acuity, thus helping children with their physical coordination and reading.

One condition that may result from or is definitely exacerbated by fatty acid deficiency, especially a deficiency of the Omega-3 fatty acids EPA and DHA is Attention Deficit Disorder (ADD) and Attention Deficit Hyperactivity (ADHD). Children and Adults suffering from ADD or ADHD and dyspraxia (poor motor co-ordination or clumsiness) have been shown to improve when their diets have been supplemented with EPA and DHA from fish oils. While the Omega-3 fatty acid, alpha-linolenic Acid (ALA), found in flax seed oil can be converted into EPA, and then EPA into DHA, this conversion is usually inefficient for most people, especially people with ADD and ADHD, and fish oil supplements have been shown to be superior to flax seed oil in improving ADHD symptoms.

However, which Omega-3 DHA/EPA supplement do you use?

For optimising mental performance, it is important to have the correct ratio of EPA to DHA. My research on brain integration using direct muscle feedback has shown that the most effective ratio for optimising brain function, especially the executive functions of the frontal lobes is 3:1 DHA to EPA. Almost all fatty acids supplements available today, and all salmon oil, have considerably more EPA than DHA, the reverse of what the brain needs to support optimal performance. Having an excess of EPA relative to DHA may slow neural transmission, thus reducing the degree of brain integration.

Only high quality tuna oil has the correct ratio of DHA to EPA, and thus supports optimum mental performance. It is important today to be sure the tuna oil is of the highest quality, as the cheaper fish oils are often contaminated with various toxins, especially PCBs and mercury. Because tuna are at the top of the food chain, miniscule amounts of these toxins are concentrated thousands of times in top level predators like tuna. So, unless the oil is cleaned by distillation or other processes, the contaminants may negatively affect your memory and neuronal function – the very thing you are trying to support.

Key Concept: Types and Ratios of Fatty Acids to Optimise Mental Performance

The brain is approximately 60% fatty acids by weight, with fatty acids important in the structure, stability and function of neuronal membranes controlling the speed and efficiency of nerve conduction. Unsaturated Omega-6 fatty acids like Arachidonic Acid and other saturated fatty acids are important in neuronal membrane structure, while the unsaturated Omega-3 fatty acids like DHA and EPA primarily control its function. DHA and EPA are conditional essential fatty acids that must come from the diet for most people, and the correct ratio of DHA to EPA is important to maximize brain function.

ROLE OF VITAMINS AND MINERALS IN OPTIMISING MENTAL PERFORMANCE

While virtually all vitamins and minerals are used in some aspect of brain function, the primary vitamins and minerals required to sustain optimum mental function fall into three groups:

1. The vitamins and minerals needed to maintain effective neurotransmitter (NT) and fatty acid (FA) synthesis to maintain adequate NT and FA levels required to support integrated brain function;
2. The vitamins and minerals needed to produce abundant energy supplies, captured as ATP, that are necessary to maintain frontal lobe function, especially when you are under stress;

3. The vitamins and minerals needed to protect the brain against free radical oxidative damage.

Vitamins and Minerals: Role in Catalyzing NT and FA Synthesis and Energy Production in the Brain

As discussed above, specific vitamin co-enzymes and mineral co-factors are required to optimise both amino and fatty acids synthesis which, if deficient, reduces levels of NT's and FA's required to maintain brain integration. Likewise, specific vitamins and minerals catalise the production of ATP, the energy needed to run your thinking and memory.

The primary vitamins required for energy production plus NT and FA synthesis are the B-group of vitamins and a matrix of micro-minerals, such as iron, zinc, magnesium and calcium. The B-group vitamins are essential nutrients with certain characteristics in common. They are all water-soluble and found in similar food sources, such as brewer's yeast, animal meat and whole-grained cereal.

Because they are soluble in water, B-vitamins are difficult to store and deficiencies can occur within weeks if intake is reduced. This is one of the reasons B-vitamins are very commonly deficient, with vitamin B deficiency 300% more common in people who do not take vitamin supplements.

B-vitamins are also interlinked by function with one B-vitamin supporting the function of another, or required in the synthesis of another. It is rare therefore to see isolated deficiency of a single B-vitamin, except B_{12} that are made by the bacteria of your gut. This is also the reason the B-vitamins should not be taken individually, but rather, together as a B-complex. (See Table 6.)

Table 6 B-Group Vitamins and the Role of each B-vitamin in the Brain

B-Vitamin	Role in Brain Function
B_1 Thiamine	Carbohydrate metabolism – energy; essential for synthesis of Acetylcholine & maintenance of myelin sheaths throughout the brain and CNS – deficiency results in loss of brain integration and polyneuritis.
B_2 Riboflavin	Major co-enzyme in carbohydrate and protein metabolism – deficiency leads to blurred vision. Important and powerful antioxidant.
B_3 Niacin	Major co-enzyme in energy production in brain – deficiency results in psychological disturbances. Powerful antioxidant.
B_5 Pantothenic acid	Essential for Krebs Cycle aerobic respiration, lipid, amino acid & cholesterol-steroid hormone synthesis – deficiency results in fatigue & neuro-muscular degeneration.
B_6 Pyridoxine	Essential coenzyme for amino acid and NT synthesis in brain – deficiency leads to loss of brain integration, memory problems and mood disturbances. Powerful antioxidant.
Folic acid Folate, folacin	Component of system synthesising RNA & DNA & essential for production of red blood cells – deficiency causes neuro-developmental problems, e.g. spina bifida.
Biotin	Essential coenzyme for steps in Krebs Cycle respiration & synthesis of fatty acids – deficiency leads to mental depression & fatigue.
B_{12} Cyanocobalamin	Coenzyme for red blood cell formation & steps in Krebs Cycle respiration & conversion of lipids and amino acids into glucose for energy – deficiency creates memory loss & neuro-psychiatric abnormalities.

Other vitamins like vitamin C and a whole matrix of minerals are involved in NT and FA synthesis. Noradrenalin, dopamine and serotonin synthesis are overtly dependent upon adequate levels of vitamins B_6 and C plus the minerals iron, zinc and magnesium, and covertly dependent upon many other nutrients. Deficiencies of any of these nutrients then can reduce NT and FA production compromising our brain integration and thus reduce our mental performance.

Free Radicals: Renegades out of Control

The other major function of vitamins and certain minerals is to scavenge 'free radical' molecules. Free radicals are molecules 'broken' during active metabolism leaving them with an unpaired electron, and the brain is one of the most metabolically active sites in the whole body.

The presence of this unpaired electron sends free radicals on a fanatic quest to find another electron at any cost. So strong is their lust for the 'missing' electron that they will literally 'tear apart' other molecules like DNA, membrane fatty acids and both structural and enzyme proteins to satisfy this need. This theft now leaves the damaged molecule with an unpaired electron often initiating a chain-reaction of molecular destruction. (See Fig. 17.)

The result is free radical damage called oxidative damage, and hence, the need for cells to have 'anti-oxidants' to protect themselves from this damage. Anti-oxidant molecules like vitamins B_1, B_3, B_5, C and E and the minerals lithium, zinc and selenium have the capacity to quench these free radicals and are often called free radical scavengers. Having a sufficient supply of these protective molecules is critical to

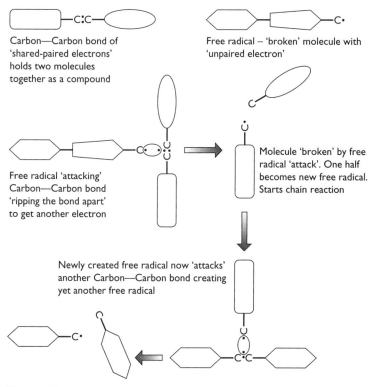

Figure 17 A Schematic of a Free Radical and Free Radical Damage.

normal brain function and guards against damage to the hippocampus, the short-term memory centre, and other brain structures!

Antioxidants are the brain's first defense against free radicals and when in sufficient supply, intercept free radicals preventing them from damaging delicate molecular structures. When antioxidants are in short supply, free radical damage may lead to cell death.

Indeed, one of the major causes of damage following a stroke is brain cell death resulting from the massive release of free radicals with not enough free radical scavengers to mop them up. Likewise, much of the brain damage resulting from alcoholism may result from the production of free radicals in the brain coupled with a dietary deficiency of mineral and vitamin free radical scavengers, especially the B-group vitamins.

> **Key Concept: The Role of Antioxidants is to control Free Radicals in the Brain**
>
> The brain is the most metabolically active organ in the body, and **Free Radicals** are molecules 'broken' during normal metabolism producing an un-paired electron. This 'broken' Free Radical molecule seeks another electron, often by ripping it out of another molecule like DNA, fatty acids in the cell membrane or proteins like enzymes, creating another free radical starting a cascade of destruction. **Antioxidants** are molecules that 'quench' the free radical reactions by absorbing the un-paired electron, thus protecting the surrounding molecules from damage. Because the brain produces high levels of Free Radicals it needs high levels of Antioxidants to prevent damage to its vital structures and molecules. The Vitamins B_2, B_3, and B_6 and the micro-minerals zinc and selenium are all powerful antioxidants.

NOOTROPIC HERBS TO ENHANCE MENTAL PERFORMANCE

Nootropic herbs act to improve learning and memory, yet have no known side effects and extremely low toxicity. Two of the best-documented nootropic herbs are *Gingko biloba* and *Barcopa monniera*. *Gingko* is the oldest known species of tree dating back 300 million years, and its leaves have been used in Chinese medicine for thousands of years. *Gingko* is used to improve cerebral circulation, mental alertness, and overall brain function.

Recent studies have shown *Gingko* to significantly improve cerebral circulation, enhance memory, especially in the hippocampus (the short-term memory centre) and working memory of the frontal lobes. A number of other studies have suggested *Ginkgo* may enhance cognitive functions in general, and no negative affects have been observed even at very high doses.

Barcopa monniera is often called by its Indian name, *Brahmi*, and like *Gingko* in China, has been used in traditional Indian Ayruvedic medicine for memory enhancement and insomnia for a thousand years. Extensive studies indicate that extracts of *Barcopa* facilitate learning and improve memory. Like *Gingko, Barcopa* has no known side effects even at high doses.

Recent work, however, has focused on its free radical scavenging capacity and the protective effect of *Barcopa* on DNA damage. Barcopa's antioxidant capacity may explain, at least in part, its anti-stress, cognitive enhancing and anti-aging effects.

Ginseng is another Chinese herb that is widely used to strengthen bodily function. It is an *adaptogen* – a compound that helps the body and brain adapt to physiological and psychological stressors. Partly it does this by shutting off cortisol release. Numerous studies have shown *ginseng* heightens cognitive functions partly because of its ability to evoke a stable state of arousal without initiating fight or flight reactions (adrenalin and cortisol release).

Ginseng comes in a number of forms, *Korean, American and Siberian ginsengs*, and among the various types there are various grades of quality. Generally a good quality *Siberian ginseng* is considered best for cognitive effects.

The Problem with Herbal Supplements

The problem with all herbal supplements is indeed the quality issue, as depending upon what part of the plant is used, how it is extracted, and how it is grown; the effects of herbal supplements may vary significantly. It is therefore important to use a good quality herbal supplement, which is unfortunately generally not the cheapest available.

Currently, the herbal market is flooded with poor quality Chinese herbs that are generally very cheap compared to other sources. So it is prudent not to purchase the cheapest herbal supplement available if you want the best results.

NUTRIENTS TO OPTIMISE MENTAL PERFORMANCE

It can be seen from these discussions that the brain needs a vast array of nutrients if it is to achieve optimal mental performance. Table 7 opposite lists nutrients suggested from studies in the scientific literature and books by bestselling authors, such as *Brain Longevity, Your Miracle Brain, Mind Boosting Secrets. Natural Supplements to Enhance your Mind, Memory and Mood,* and *Optimum Nutrition for the Mind* to provide optimal nutritional support for the brain.

Table 7 Nutrients to Maintain Optimum Mental Performance

Nutrient	Recommended Daily Dosage
Vitamins:	
Vitamin A	5,000–25,000 IU
Vitamin B_1	50–1000mg
Vitamin B_3	30–200mg
Vitamin B_5	100–1000mg
Vitamin B_6	50–200mg
Vitamin B_{12}	100–1000mcg
Folic Acid	200–400mcg
Vitamin C	500–3000mg
Vitamin E	100–800mg
Minerals:	
Magnesium	100–300mg
Calcium	50–250mg
Selenium	50–100mg
Iron	10–30mg
Zinc#	30–50mg#
Lithium	10–30mg
Amino Acids:	
Glutamine	100–250mg
Phenylalanine	100–500mg
Tyrosine	100–1000mg
Tryptophan*	100–500mg*

#There should be a number of different types of zinc to maximize uptake & utilisation in the many roles zinc plays both within the brain and the body.

* Tryptophan was removed from the market ~10 years ago and now is only available at low doses in many countries.

Fatty Acids:	
DHA/EPA (3:1 ratio)	250–1000mg
Herbs:	
Gingko biloba	50–500mg
Barcopa (Brahmi)	20–300mg
Ginseng	500–1500mg

Clearly, the dosages of the nutrients listed in the Table 7 below vary over a wide range, from doses just above the RDA for nutrients like iron to megadoses many times above the RDA. For instance, the suggested dose of vitamin C is 5 to 30 times the current RDA of 75 mg, while the suggested dose of B_6 is 25 to 100 times the RDA of only 2 mg. However, the RDA's are often hundreds or more times below toxic levels. For instance, Vitamins B_3 and B_6 have been found to be deficient in both disturbed children with behavioural problems and adult schizophrenics. This appears to result at least in part from malabsorption from a diet that may contain adequate amounts of these nutrients for people with normal absorption, again, the genetic factor.

One patient required B_6 supplementation of 1600 mg per day before measurable levels were detected in their blood. This is more than 800 times the normal daily requirement of 2 mg per day for adult men and women. However, B_6 toxicity is not generally observed until more than 2000 mgs per day are taken by people with normal absorption, and 200 mg per day has been shown to be helpful in the treatment of asthma, and high does improved memory of men between 54 and 80 years of age. Most nutritional texts and the research literature would suggest B_6 supplementation with between a minimum of 10 and a maximum of 200 mg per day, that's 5 to 100 times the current RDA.

The same was true for vitamin B_3, with some people may requiring between 850 and 1800 mg of B_3 per day, far in excess of the normal daily requirement of 12 to 18 mg, and yet still have low B_3 levels in their blood. One schizophrenic taking over 3000 mg per day was found to have only borderline low B_3 levels in his blood. A similar situation was observed with children with behavioural problems dependent on B_3. As soon as a placebo was substituted for the nutrient, all children relapsed in their behaviour within 30 days, but recovered again when B_3 supplementation was restored. It thus appears that for some people with poor absorption of B-group vitamins, commonly people displaying behavioural problems, supplementation is almost essential for their optimum cognitive performance

It should be noted that Tryptophan was used safely for many years to treat depression and sleep disorders but was banned about ten years ago following the death of several people. However, even though the cause of these deaths was clearly shown to be bacterial toxins that contaminated the tryptophan during manufacture soon after the incident, it was not allowed for sale again until recently, and many countries still only permit very low doses. So, while the levels suggested in Table 7 are higher than allowed in these countries, they are certainly well below levels shown or known to be toxic.

While some of these doses sound 'high', they are no where near levels that may be toxic, which for most vitamins and minerals are generally 10 to 100s of times higher than therapeutic doses suggested, and then only if taken for extended periods of time. Several studies of thousands of people having various illnesses and diseases have demonstrated time and again, that the doses of vitamins and minerals demonstrating therapeutic effects are 5 to 50 times greater than the RDAs, especially in cases of chronic deficiency.

Only people taking very high dosages of vitamins or minerals, and then over long periods of time need be concerned about vitamin or mineral toxicity. But this can also be said of any food, even salt or sugar.

While the list of nutrients to support mental function in Table 7 may seem to the average person – a lot!, it is really only the basic nutrients absolutely needed for enhanced brain function. There are a host of other nutrients like lecithin (to provide phosphatidyl choline); the 'other' amino acids: acetyl L-carnitine, L-taurine, L-methionine involved in other aspects of brain metabolism like detoxification; DMAE (dimethylaminoethanol – a source of choline for the NT, acetylcholine); Coenzyme Q-10 (a powerful antioxidant and critical molecule in cellular respiration); and a number of other antioxidants like 'green juices' and green tea plus various other nootropic herbs like Ching Chun Bao that have all been shown to help brain function.

While other additional nutrients and herbs may be of assistance, the nutrients and nootropic herbs I have selected in Table 7. have all been widely acknowledged as enhancing various aspects of brain function and mental performance such as enhancing memory and will provide a solid nutritional foundation upon which to build long-term optimum mental function.

SYNERGY: THE MISSING PIECE OF THE NUTRIENT PUZZLE

What is often overlooked in the suggestions for each of the brain enhancing nutrients above is – 'How much do you need when a balanced formula of all relevant nutrients is taken together, including vitamins, minerals, amino acids, fatty acids and nootropic herbs?' While multivitamin and multi-mineral supplements are common, these combined supplements seldom contain therapeutic amounts of amino acids or fatty acids and with few exceptions, no nootropic herbs.

Most Western studies of supplements only include a single, or at most, two or three nutrients and thus the amounts found effective and/or recommended only reflect the effects of that specific nutrient. Because nutrients work together synergistically, with one nutrient supporting the function of another nutrient, the amounts of individual nutrients required is often decreased significantly if they are all taken together as a single matrix.

If you really want to enhance your brain function, how do you take 23 different vitamins, mineral, amino acid, fatty acid and herbal supplements all at the same time and in the correct quantities and ratios? This is the role of nutriceuticals designed specifically to enhance mental performance.

Chapter 9

NUTRICEUTICALS FOR OPTIMIZING MENTAL PERFORMANCE

INTRODUCTION

When you look at the sheer number of nutrients required to support optimum mental performance, it seems unlikely that most people would ever take all of these supplements, especially as they need to be taken all at the same time. Add to this the need to take additional nootropic herbs to further enhance brain function, and the practicality is further reduced.

But if you want to improve or optimise your mental performance, what should you do?

Nutriceuticals were developed exactly for this reason – that is, to enable the average person to enhance specific functions (such as mental performance) with nutrition, but in a way that was both effective and convenient.

Instead of each person trying to read all of the literature, much of it conflicting, then trying to figure out exactly what they should take, followed by purchasing all of the individual supplements and trying to figure out exactly how much of each supplement and in what form it is needed, nutriceuticals have been developed that do all of this complex work for you.

Nutriceuticals are a whole new concept in nutrition. Whereas pharmaceuticals are drugs developed to treat specific diseases and conditions, nutriceuticals are nutritional formulas designed to provide all of the nutrients necessary to optimise particular functions – for instance, one easy-to-take supplement to optimise mental performance.

Development of Nutriceutical Formulas for Enhanced Mental Performance

In recent years, a number of companies have begun to produce nutritional formulas to enhance brain function, as the role of nutrients in enhancing brain function has become better known. Today there are a number of nutritional formulas on the market that claim to help brain function thus increasing your memory, focus and/or attention such as Attend, Focus Factor, Learning Factor, Sense of Mind, Brain Link and ThinkingAdvantage. Brain Link even claims to support the right and the left hemisphere integration and to enhance brain function, while ThinkingAdvantage was designed by direct biofeedback to support both short-term and working memory functions at the same time balancing the Amygdala and survival systems (See Appendix 1).

The rapid growth in the number of these products is a result of the public becoming ever more informed about brain function in general, and the vital role nutrition plays in our health and mental function. There are even books on Smart Nutrition and many articles in popular magazines emphasizing the importance of nutrition for memory and enhanced brain performance.

While many of these products claim to enhance or improve mental function and memory, the following questions need to be asked:

1. How were these formulas developed?;
2. How were the different components selected?;
3. How were the forms of each nutrient chosen?; and
4. Upon what basis were the ratios of each nutrient relative to the other nutrients in the formula determined?

From the discussions in the previous chapters, these are all important questions that are seldom, if ever, asked.

How have these products been developed?

In most cases, an individual or a company has a concept for a new supplement to provide support to a particular system or function, for instance brain function. They usually first look to see what other manufacturers have put in their products, and then go to the literature to see what has been published about nutrients that may enhance brain function. After comparing the composition of other products and reviewing the literature, they then choose a variety of nutrients and combine them into their new product.

More recently, direct biofeedback has been used to look at both the nutrients needed, and the best forms to create synergistic function, including energetic components such as homeopathics.

Upon what basis were the specific nutrients in these products chosen?

Some companies do this by limiting their composition to those nutrients that have been shown to support specific functions like the use of *Gingko* to support memory. Then a single or, at most, a few components are selected that appear to be very important to brain function – what I call the 'Big Guns' approach. The other common approach is what I call the 'Kitchen Sink' approach, in which a little of everything that has ever been shown or suggested to support any brain function is put together.

The Problem with the 'Big Guns' approach

The 'Big Guns' approach of using a single component or a limited number of components in the formula *often makes a false assumption*: this is that all of the ancillary nutrients necessary to effectively utilize the 'Big Guns' are actually available in the brains of the people taking their product. As extensively discussed in the preceding chapters, most people suffer marginal nutritional deficiencies in one or more vital nutrients of the complex ancillary nutrient matrix required for the 'Big Guns' to fire.

Thus, very often important supporting characters needed to load, aim and fire the gun are absent or over-taxed and therefore cannot do their job effectively. Without a complete nutrient matrix to support the biochemical reactions necessary to effectively utilise these important nutrients, these formulas often prove to be ineffective for many people.

For instance, more than 50% of Americans are iron deficient, and iron is a major co-factor required for the production of almost all of the neurotransmitters in the brain. Thus, Iron deficiency alone can reduce production of essential neurotransmitters and fatty acids even in the presence of high concentrations of precursor (the 'Big Guns') provided by formulas that *should, in theory,* sustain optimum brain function. Or, what if you are chronically zinc deficient? Since zinc-based enzymes largely control frontal lobe and memory functions, just the deficiency of this one nutrient can undermine the effectiveness of the whole supplement, even though it contains other positive ingredients.

The Problem with the 'Kitchen Sink' approach

The 'kitchen sink' approach produces products that technically *should* work, as they contain everything the brain needs to work effectively, but usually produce only limited effects for three reasons. First, while individual nutrients may show effect when used singly, when in combination with many other nutrients, they may actually be less effective because of competition for uptake by receptors and transporters that need to get these nutrients into the cells. For instance, the brain needs the essential amino acids, phenylalanine, tyrosine and tryptophan to make the major modulatory NT's to maintain brain integration. But many supplements commonly contain a wide array of up to 20 amino acids. Since all amino acids use the same transporters to enter the brain, the presence of large numbers of competing amino acids in the formula actually reduces the concentrations of these essential amino acids entering the brain, lowering the levels of the important modulatory NT's.

Secondly, because of the extremely large number of ingredients in these formulas, each ingredient is usually not present at the therapeutic concentrations needed to be effective. And having only 10 milligrams of an important nutrient when you need 50 milligrams for optimum function, greatly reduces the effect of the supplementation.

And lastly, while there are many individual components, the degree of synergistic interaction between the various components is unknown, and often limited by having the wrong form of the nutrient in the formula. And, in biochemistry of the body, form is everything because the shape determines where, how and how effectively a nutrient can be used. For example, while ionic zinc may have very limited effect in raising intracellular zinc levels and a single form of chelated zinc may take many months to change a chronic deficiency, a complex organically bound zinc supplement may rapidly eliminate the same chronic deficiency.

THE COST FACTOR: SELECTION OF INGREDIENTS

Once the list of ingredients has been decided upon, then the relative cost of each component and of the different forms of each component, is evaluated. For instance, Calcium may be correctly chosen as an important component, but do they choose Calcium Carbonate, which is very cheap but poorly absorbed, or do they choose a much more expensive chelated Calcium that is highly absorbed and rapidly utilized in the body?

Check the labels and you will often find the first choice, which, while less expensive for the manufacturer, is bought at the cost of actual effect in the person taking the supplement.

In addition, many manufacturers, like many people, think, 'zinc is zinc', or 'calcium is calcium' and what form of the nutrient is used doesn't really matter very much. As discussed Chapter 4 with zinc supplementation, the actual form of the nutritional supplement determines how, where and how much will be used in the body. So zinc supplements with zinc sulphate, even though technically containing a lot of zinc, may poorly support mental performance because so little of it is actually transported into the brain.

Upon what basis were the specific concentrations of each ingredient determined?

Once having selected the components, how *much* of each ingredient is included in the product? Commonly, standard nutritional guidelines (most often the Recommended Dietary Allowances or RDA) are used to determine concentrations. Nutrient concentrations of 25% to 100% of the RDAs are generally considered adequate, just check the nutritional information on any cereal box or vitamin or mineral supplement. However, research shows that RDAs do not represent levels to support optimal function, but rather only the minimal levels needed not to become overtly ill.

This is an important point because, to have maximum effect, an effective product must often have far greater concentrations of a specific nutrient than the RDA. Based on the more recent scientific literature, Therapeutic Dosage Range (TDR) of many nutrients are 5 to many times the RDA.

How do the makers of these products know if they have chosen effective therapeutic doses that will maximize brain function? Many manufacturers don't because most of the concentrations were merely based on RDA values or simply guessed.

How do they know that the combination of components they have chosen is synergistic (work together for greater effect), neutral or antagonistic to each other? Again, they probably don't know because of the overheads in testing for these factors! It is very expensive and difficult to test each combination, as you would literally have to trial each of the many different combinations for several months to see which one is best, and this is prohibitively expensive.

While many manufacturers claim each of their products has the 'right' combination of ingredients based on the latest scientific findings, and that these products work for everyone, this is often not the case for all of the reasons above.

Who needs these Formulas?

As we have seen, there is a wide range of people who might well benefit from supplementing their diet with high quality, well-balanced nutriceuticals for reasons varying from lifestyle, dietary choices to genetics.

- People suffering from the various forms of stress discussed in Chapter 6. For example high-powered professionals working long hours with tight schedules, pupils studying for exams, and housewives 'juggling' busy day to day household activities.
- People whose genes mean they do not absorb and process zinc and/or other nutrients well, or who produce 'slow' versions of important enzymes, and thus cannot support normal levels of brain activity and function with normal dietary levels of nutrients.
- People who do not eat a balanced diet due to such factors as poor quality food (due to depletion of nutrients in the soil and loss of nutrients in storage and handling) or poor choices of food (such as fast foods, foods containing too much sugar, and highly processed foods). For example,

people who try to eat properly but the food just doesn't contain enough of the essential nutrients, workers who don't take the time to eat properly, or children who only eat a limited choice of foods.

With stress implicated as an important causative factor in over 50% of chronic illnesses and the widespread prevalence of marginal nutritional deficiencies in most Western populations, it would seem that an increase in the levels of nutrition by supplementation and a reduction in stress levels would be of great benefit to a large number of people in today's society.

Factors to consider when Supplementing for Optimum Mental Performance

Clearly while many products make claims about their ability to support enhanced brain function, can they deliver what they promise? More importantly, how can you decide which one might work well for you?

No nutritional product can work for everyone, as individual needs vary enormously, and deep-seated emotional problems can clearly interfere with optimal mental performance, thus reducing response to nutritional supplementation alone. While many of the products currently on the market do work for some people some of the time, they often do not work for a very high percentage of the people most of the time for the following reasons:

1. They contain only the 'Big Guns', one or a few important nutrients, but lack an effective supporting nutritional matrix.
2. They contain many relevant nutrients, but in too low amounts to be effective, e.g. 25% of RDA.

3. They contain poorly absorbed or less effective forms of important ingredients, e.g. zinc sulphate, calcium carbonate or cheap herbs.
4. There is little synergistic interaction of the ingredients used in the product, limiting their overall effectiveness.

With the above considerations in mind, if you wish to 'feed' your brain for optimum mental performance, a good place to start would be to evaluate the nutriceuticals and nutritional formulas to enhance brain function currently available. You may wish to consider the following guidelines in this evaluation. Does the product have:

- Essential Amino Acids that are precursors to the primary Neurotransmitters in the brain?
- Essential Fatty Acids to support effective nerve conduction and neuronal membrane stability?
- The correct ratio of Omega-3 fatty acids DHA and EPA with more DHA than EPA, a 3:1 ratio being preferred?
- Adequate levels of most components – that is generally 100% of the RDA, or preferably several times the RDA for that nutrient?
- High quality ingredients and effective forms of the major components, e.g. chelated minerals, not inorganic minerals like zinc sulphate or calcium carbonate?
- High quality Nootropic Herbs like Gingko, Barcopa and/or Ginseng to enhance brain functions?
- Do the ingredients demonstrate synergistic effect?

While the last point above is difficult to know from the label alone, if the product meets the rest of the above criteria, there is a reasonable probability that it will be effective.

Why Do I Need to Take Supplements if I Eat a Balanced Diet?

Many traditional orthodox sources will say that if you eat a balanced diet containing 100% of the RDA of all the major nutrients, then nutritional supplementation is just a waste of money. But, upon what basis do they make this statement?

Certainly this statement is not based upon the scientific literature that demonstrates exactly the opposite for many people who do not possess perfect genes. Nor is it based upon a careful evaluation of the quality of much of the produce and the nutritional content of the processed foods actually consumed by many people in Western societies. And it is certainly not based upon the actual food choices many people make, otherwise the junk food industry could not be the multi-billion dollar industry it is today!

This statement is only true if you happen to meet the following criteria:

- Perfect genes so that all receptors, transporters and enzymes are fast and efficient.
- Consume an organic, nutrient-rich diet containing a full complement of nutrients with several to many times the RDA for most nutrients.
- Have a habit of making only 'good' food choices, avoiding all junk and highly processed foods.
- Live a low Stress lifestyle.

Only if you have or meet all of the above criteria would supplementation represent a waste of money.

However, with today's foods and the food choices we all make, the widespread presence of genetic variation in the ability to absorb and utilize nutrients, and the levels of stress

> **Key Concept: The Statement that Nutritional Supplementation is just a Waste of Money because all you need to be Healthy and have Optimum Brain Function is to eat a Balanced Diet is simply NOT True for most People Today**
>
> This statement is based upon a number of assumptions, all of which are scientifically known to be untrue for a large percentage of the population. This includes all of us who do not have perfect genes, who do not eat organic foods with high levels of all nutrients, who do not always make the best food choices consuming Processed, Junk and Fast Foods with various Nutrient Deficiencies, and who do not Live a Low Stress Lifestyle.

many people in modern societies live with daily, it is unlikely that many people will get all of the nutrients they need to maintain optimal mental function from their diet alone. This is especially true when you are under higher levels of stress!

What about the Cost – Isn't it Expensive?

What is expensive is largely a point of view. If to feed your brain what it needs to maintain optimal function costs less than two euros or dollars a day – 'Is that expensive?' Many people would answer – 'Yes, absolutely!'

However, if you ask the same people if they drink a Coke and have a Mars Bar a day? Or do they smoke a pack of cigarettes a day? Or do you have a cappuccino or alcoholic drink every day? Many of these same people would answer 'Yes!' to this question as well.

Do they consider the Coke and Mars Bar, or cigarettes or cappuccino or drink expensive – probably not! So whether the cost of supplementation is expensive or a reasonable price to pay for effective mental function is all in your point of view!

Glossary of Terms

Absolute Nutritional Deficiency *Absolute Nutritional Deficiency* is depletion of nutrients in the body because of their absence from the diet, often from starvation.
Also see Marginal Nutritional Deficiency.

Acetylcholine *Acetylcholine* is an excitatory neurotransmitter made from the acetyl group produced by the metabolism of glucose and the choline molecule from various sources, lecithin being one good source. Acetylcholine is important in muscle function, activating the cortex so you are 'awake', and important in hippocampal memory as well as autonomic nervous system functions. Many drugs may affect the production and release of this neurotransmitter.

Adrenalin *Adrenalin,* also called epinephrine, is a stress hormone that is part of the Fight or Flight reaction released by the adrenal medulla. Adrenalin causes the release of sugar into the blood, increases heart rate and blood pressure, increases the power of muscle contraction and inhibits frontal lobe function at higher levels.

ALA (alpha-linolenic acid) *ALA* is an *essential* long-chain omega-3 fatty acid with three double bonds that can be converted into EPA, and then into DHA. The production of DHA from ALA, however, requires sufficient nutrients as well as four fast effective enzymes therefore a deficiency of any of the ancillary nutrients or slower enzyme variants can slow production and result in DHA deficiency in the brain.

For this reason, EPA and DHA are said to be 'conditional essential' fatty acids.

Alleles *Alleles* are genes that make the same protein, but differ in the exact sequence of amino acids. Because the sequence of amino acids determines the shape of a protein, each Allele produces a slightly different shaped version of the protein, and the shape determines its function. Evolution selected for fast, efficient enzymes, transporter and receptor molecules, but due to mutation some Alleles produces slower, less efficient enzymes, transporter and receptor molecules.

Amino Acids *Amino acids* are chains of carbon atoms with an Amine group (NH_2-ammonia) at one end and a Carboxylic Acid (COOH) group on the other end. Some amino acids have 'side-chains' or other 'functional' groups attached along the chain of carbon atoms. Chains of amino acids form the proteins for our bodies and act as precursor molecules for neurotransmitters (NTs) that create brain integration. Some of these precursors need to be present in abundant supply in order to maintain NT levels, especially when under stress.

The precursor amino acids for the major Neurotransmitters in the brain are phenylalanine, tyrosine, tryptophan, glutamine and glycine, with phenylalanine, tyrosine and tryptophan only available through the diet, hence essential Amino acids.

Ancillary nutrient matrix *Ancillary nutrient matrix* is a specific formula of ancillary nutrients which, together with other molecules, optimize enzymatic action and the transport and reception of important molecules maintaining brain function. An important factor in maintaining optimum mental function is the synergy between different nutrients, with one nutrient supporting and optimizing the function of another. It is therefore important that the body has a full supply of these nutrients at all times.

GLOSSARY OF TERMS

Ancillary nutrients *Ancillary nutrients* required in order for other nutrients to be utilised effectively. For instance, the Krebs Cycle producing cellular energy in the form of ATP overtly requires vitamins B_2 & B_3, and the minerals iron, copper and sulphur, but also up to 40 other nutrients to maintain the complex biochemistry involved in this process. Likewise, production of most of the major brain neurotransmitters requires vitamin B_6 and the minerals iron, zinc magnesium and calcium in ample supply to optimise their production.

Anti-oxidants *Anti-oxidants* are substances that protect against free-radical damage. Some are termed nutritional antioxidants such as Vitamins B_2, B_3, B_5, C and E and the minerals lithium, zinc and selenium, as well as glutathione that protect cells from free radical damage. Others are termed enzymatic antioxidants such as Super-Oxide-Dismutase (SOD). As soon as a free-radical is produced either in normal metabolism or from a toxin, it should immediately react with an antioxidant turning off its terrible destructive power.

Because many free radicals are produced during metabolism and the brain is the most metabolically active site in the body, it is critical for effective brain function to maintain a sufficient supply of these protective molecules to guard against damage to the brain.

Arachidonic Acid (AA) *Arachidonic acid or AA* is a long-chain Omega-6 fatty acid with twenty carbon atoms and four double bonds. It is the brain's principal Omega-6 fatty acid, where it is a major structural constituent of neuronal membranes. AA can be converted into powerful inflammatory Prostaglandins (PGE 2). However, Alpha-linolenic acid, EPA and DHA can counter these inflammatory effects of arachidonic acid.

ATP *ATP* is the primary energy molecule or energy currency of the body. It is made from Adenosine nucleotide (one of the base pairs forming DNA) with three phosphate groups attached by high-energy bonds, and is thus called Adenosine Triphosphate or ATP for short. The splitting of ATP to ADP releases the energy used to run all of your cellular activities from muscle contraction to thinking.

The production of ATP is heavily dependent upon dietary nutrients, including fatty acids, and the absence of sufficient quantities of certain vitamins and minerals in the body reduces the amount of ATP that is produced, slowing our thinking.

Axon *Axons* are the extensions of the nerve cell that carry the nerve impulse away from the cell body, which has integrated all of the inputs usually from many Dendrites. If the signals arriving at the dendrites are more excitatory than inhibitory, then the cell body integrates these inputs and creates a wave of depolarization, the nerve impulse, which then travels to the end of the axon where there is a synapse with the next neuron. If enough Neurotransmitter is released, and activates receptors on the dendrite of the next neuron, this may 'fire' the next neuron by creating a new wave of depolarization to travel down its axon.

B-Complex Vitamins A *complex of B-vitamins*: B_1, B_2, B_3, B_5, B_6, B_{12}, Folic Acid and Biotin are needed for optimum brain function. Because B-vitamins are interlinked by function with one B-vitamin supporting the function of another, or required in the synthesis of another, it is rare to see isolated deficiency of a single B-vitamin, (except B_{12} that is not found in vegetables and requires Intrinsic Factor made by bacteria in the gut to be absorbed), therefore B-vitamins should not be taken individually, but rather, together as a B-complex. The B-vitamin Biotin, in particular, is necessary for the metabolism of amino and essential fatty acids and is important for brain function.

Vitamin B_6 is essential for the production of several major brain neurotransmitters such as Dopamine, Serotonin and Noradrenalin that control our moods, learning and behaviour, while B_2 & B_3 are essential for energy production in the brain.

GLOSSARY OF TERMS

Brain Integration   *Brain Integration* is a concept developed in the LEAP® Program based upon neural processing in the brain. Because thinking and memory rely upon functions widely distributed in the brain, these functions depend upon maintenance of the precise timing and synchrony of neural flows within the brain to provide integrated function. Thus loss of this timing and synchrony results in the loss of mental functions, or loss of Brain Integration.

At its most basic level, brain integration is a biochemical phenomenon as it relies on nutrients (essential amino acids and fatty acids and ancillary nutrients), needed to maintain the production of sufficient neurotransmitters and stable membranes to transmit the nerve impulse from one neuron to the next as well the production of receptor and transporter molecules.

Also the brain uses enormous amounts of energy to function, and thus requires numerous nutrients to provide this energy. When nutrients are in short supply the energy production of the brain is reduced, resulting in a shift from energy expensive thinking to energy conservative survival reactions, and loss of integrated brain function or loss of Brain Integration.

Calcium   *Calcium* is a macro-mineral that is found in high concentrations in bone, plays a major role in muscle contraction, and is a major regulatory mineral controlling many cellular reactions.

There are different types of calcium: Calcium Carbonate is a common form found in many supplements (cheap to make, but very poorly absorbed), while the more expensive chelated forms of Calcium that have higher rates of absorption and utilization are less often used.

Chelation, Chelates   *Chelation* is a chemical process by which a molecule with a 'pincer-like' structure holds another atom, usually a charged metal ion such as zinc in this pincer-like structure, and then removes the *chelated* metal from the body normally by excretion in the urine.

Chelation comes from the Greek '*chela*' or pincer.

Choline *Choline* is a nutrient in the B-group that the brain uses to make Acetylcholine. Eggs and lecithin form soy are two important dietary sources and choline precursors or sources are sometimes used as a supplement to enhance mental functions, especially memory as effective hippocampal function, the short-term memory centre, is highly dependent upon sufficient levels of Acetylcholine to function properly.

Co-Enzyme *Co-enzymes* are vitamins needed to activate enzyme action by joining with the inactive form of the enzyme, the apoenzyme, the combination forming the active form of the enzyme. The levels of vitamins thus control enzymatic action. For example, B_6 is an important co-enzyme for the reactions producing most of the major neurotransmitters in the brain, and thus, B_6 deficiency can cause loss of Brain Integration.

Coenzyme Q10 *Coenzyme Q10,* also know as ubiqinone or simply CoQ10, is a nutrient that serves two primary functions: It is an antioxidant and functions as a key nutrient in mitochondrial generation of ATP via the Krebs cellular respiration cycle. Thus, CoQ10 is very important in maintaining brain integration as it both protects the brain against free-radical damage and produces the energy the brain needs to function.

Co-Factors *Co-factors* are one of the important classes of microminerals that, along with vitamins, control enzyme action. Mineral co-factors change the shape of the inactive apoenzyme into its active form thus optimising both neurotransmitter and fatty acids synthesis.

For example, iron and zinc are major *co-factors* for the production of almost all of the neurotransmitters in the brain, and zinc is an important *co-factor* in many enzyme systems supporting both body and brain function, especially thinking and memory.

GLOSSARY OF TERMS

Deficiency of vitamin co-enzymes or mineral *co-factors* will significantly reduce enzyme activity, a major reason people feel so tired when they are deficient in B-vitamins or the minerals iron, zinc and copper.

Cortisol *Cortisol* is a stress hormone, part of the Fight or Flight reaction, released by the adrenal cortex that increases the release of sugar into the blood, suppresses the immune system to divert energy to survival functions, and blocks the 'now time' memory of the hippocampus (your short-term memory centre) in favour of reactive memory of what you did last time to survive. Chronic high levels of cortisol from prolonged traumatic stress can damage the hippocampus and lead to chronic poor health due to immune suppression. It is therefore important to reduce your stress levels in order to decrease levels of Cortisol.

DHA *Docosahexaenoic acid (DHA)* is a long-chain Omega-3 fatty acid with six double bonds, and as such, is the most flexible fatty acid in the body. Due to this flexibility, it is able to provide higher rates of membrane activity than any other fatty acid, and is thus, the primary fatty acid in the retina of the eye and at the axonal end bulbs where neurotransmitters are released and taken back up. DHA is the primary fatty acid in the brain that controls neural signaling and receptor activity. The brain is 60% fatty acid, 25% of which is DHA, whereas other tissues of the body have much less DHA.

Technically DHA can be made from the dietary fatty acid alpha-linolenic acid (ALA), but because of the 5-enzymatic steps involved conversion rates are usually low. This makes DHA a 'conditional' essential fatty acid, which means for most people it must be provided in adequate levels in their diet.

Dendrite *Dendrites* are fine extensions of the nerve cell body that project out in many directions usually being highly branched. These are the primary sites where the axons of other neurons end as a synapse between the axon on the upstream side and the dendrite on the downstream of neural flow. (See Synapse) The dendrites may have *Dendritic Spines*, like twigs projecting from a tree branch, increasing the possible number of synaptic connections to between 1000 and 20,000 synapses per individual neuron in the brain. The Purkinje cells of the cerebellum have highly elaborate *Dendritic Trees* with over a 100,000 synapses with the axons of other neurons.

Dopamine *Dopamine* (like Noradrenalin) is an important neurotransmitter that controls how 'good' we feel, and how well we handle stress. It plays a major role in motivating you to seek reward, but is also important in control of movement. Drugs such as alcohol, marijuana, cocaine and heroin cause excess amounts of dopamine to be released into the basal forebrain Nucleus Accumbens, a primary reward centre giving that 'high' feeling.

Dopamine released by the Substantia nigra of the brainstem also smoothes out muscle action but, when these dopamine secreting neurons begin to die off, it causes the shaking and tremors of Parkinson's Disease.

Phenylalanine and Tyrosine are pre-cursor essential amino acids required for the production of Dopamine along with a nutrient matrix of Vitamin B_6, magnesium, iron, and zinc.

DV *Daily Value.* This is an officially recommended amount of a nutrient you need to take on a daily basis not to develop nutritional deficiencies. It is similar to the RDA (Recommended Dietary Allowance) and like the RDA, it is not the amount of nutrient you need to be healthy, but the amount you need not to suffer a nutritional deficiency disease, and does not indicate which nutrients should be taken together, i.e. synergy of the nutrient matrix.

EPA *Eicosapentaenoic acid (EPA)* is a long-chain Omega-3 fatty acid, one of the most important omega-3 fatty acids in the body that is especially important for the function of the circulatory system. It is also responsible for maintaining stable neuronal membranes necessary for optimum brain function, for cardiovascular health and for reducing the incidence of inflammatory diseases such as arthritis. The enzyme Delta-6-desaturase can convert EPA to DHA, but too much EPA relative to DHA can decrease neuronal membrane efficiency in the retina and brain, therefore a balance between EPA and DHA is required.

Enzymes *Enzymes* are proteins in the body that act as catalysts to make a chemical reactions proceed at normal body temperature. Enzymes make it possible for you to 'burn' glucose or other carbohydrates by greatly reducing the amount of energy needed to 'break' chemical bonds, or 'form' new chemical bonds joining two or more molecules together like the acetyl and choline molecules that make up the neurotransmitter Acetylcholine. For instance, the Krebs Cycle is a series of enzymatic steps that converts glucose into carbon dioxide and water while providing a controlled release of energy that is then converted to ATP using yet other enzymes. Enzymes are involved in making molecules from their precursor molecules, called *anabolism*, or taking molecules apart, called *catabolism*. Delta-6-desaturase in an enzyme that creates more double bonds in fatty acids, so it is important in converting ALA to EPA and DHA.

Essential nutrients *Essential nutrients* are those nutrients you can only get from your diet, as they cannot be made from other nutrients as Non-Essential nutrients can. Also see *Non-Essential nutrients*.

Excitatory Neurotransmitters (NT) *Excitatory Neurotransmitters* decrease the polarity or charge of the nerve membrane that once it reaches −59 millivolt, triggers a wave of depolarisation to roll down the axon to the end bulb where neurotransmitter is released. This rolling wave of depolarisation is called a *nerve impulse*. Once several nerve impulses have released enough neurotransmitter at the synapse, the next neuron will be depolarised and pass this nerve impulse down its axon to the next neuron and so on. That is unless an inhibitory neuronal end bulb from another axon, often from another part of the brain, releases enough Inhibitory Neurotransmitter to 'stop' the wave of depolarisation created by the Excitatory Neurotransmitter, 'turning off' this message.

The balance between Excitatory NTs and Inhibitory NTs determines whether the message is passed on or stops! Glutamate, Glutamic acid and Acetylcholine are three important excitatory neurotransmitters in the brain.

Fatty Acids (FA) *Fatty acids* are long chains of saturated or unsaturated molecules that make up the membranes of our cells and come in 'essential' and 'non-essential' forms. Fatty acids play a major role in the structure and function of the brain and other body functions and are the precursors to important steroid hormones such as estrogen, testosterone and cortisol.

Fight or Flight *Fight or Flight* is an integrated set of reactions run by the brainstem for physical or psycho-emotional survival triggered by certain types of stress. When the survival centres in the brain perceive 'threat or danger', they initiate a whole cascade of physiological events that prepare the body to 'Fight or Flight'. Among other processes, there is an immediate release of adrenalin and cortisol from the adrenal glands. This release of these stress hormones increases the heart rate and blood pressure, releases glucose into the blood, while at the same time switching off conscious memory and frontal lobe thinking in favour of reactive memory in order to survive.

Free-Radicals A *Free-Radical* is a highly reactive molecule, atom or molecular fragment with a free or unpaired electron. Free radicals are produced in many different ways such as from:

normal metabolism, ultraviolet radiation, nuclear radiation, and the breakdown of various molecules in the body such as fatty acids. The presence of this unpaired electron will literally 'tear' other molecules such as DNA, membrane fatty acids and proteins apart to get the other electron it needs for stability. This theft of an electron often initiates a chain-reaction of molecular destruction by producing more free-radicals, called a free-radical reaction.

Free-Radical Reactions *Free-Radical Reactions* take place in the body countless times each day, made worse by emotional stress, ultraviolet light, toxic substances and cigarette smoke. Free-radicals have been linked to more than 60 diseases, mostly the diseases of aging, cardiovascular disease and cancer. Fortunately, *antioxidants* quench or attach to *free-radicals* turning off free radical-reactions. Thus having a sufficient supply of anti-oxidants is important to ensure the brain is protected from free radical damage. (See Antioxidants)

Functional Groups *Functional groups* are various biochemical elements that attach to amino acids, proteins, sugars and fatty acids molecules that change their 'function'. These may be simple inorganic molecules like the Hydroxyl group ($-OH^-$), or complex chains of amino acids or cyclical aromatic groups like the benzene ring ( )

Functional Groups may attach to amino acids and fatty acids and change their activity or function. An example, the production of the neurotransmitter noradrenalin occurs by merely adding two simple functional groups, two hydroxyl ions, to the precursor amino acid tyrosine, and noradrenalin is a neurotransmitter, tyrosine is not.

GABA *GABA, or gamma aminobutyric acid*, is an amino acid that is one of the primary inhibitory neurotransmitters in the brain. While GABA can be overtly inhibitory, turning off certain nerve impulses, one of the primary functions of GABA inhibition is to synchronise neural flows in the brain. About 40% of the neurons in the frontal lobes release GABA, as one of the primary functions of the frontal lobes is to 'inhibit' competing inputs to synchronise brain function.

Ginkgo biloba *Gingko biloba* is one of the oldest types of trees in the world, being present at the time of the dinosaurs, and extracts of the leaves and fruit of this tree have been used for millennia in China to help support brain function. We now know the active ingredients are a group of chemicals called gingosides and that these are potent stimulators of cerebral circulation. A number of studies have shown that supplementing with Ginkgo extracts can improve memory and general mental function of older people, even those with Alzheimer's Disease.

Glutathione *Glutathione* is a sulphur-containing nutritional antioxidant that protects the body and brain from free-radical damage, especially in the liver and brain. Depletion of glutathione stores in the brain appears in some brain disorders and in normal aging.

Glucose-Tolerance Factor The *Glucose-Tolerance Factor* is necessary for proper glucose metabolism as it permits insulin to activate cellular uptake of glucose from the blood, providing the glucose needed to run cellular respiration and make the energy molecule ATP. This metabolism is especially important as the brain uses more glucose gram-for-gram than any other tissue in the body.

When the brain is working intensely, even without stress, your brain may consume more than 50% of all the glucose in your body. When you add stress, it consumes even higher levels.

Hippocampus The *Hippocampus* is the short-term memory centre in the brain. It has a unique wiring that creates self-sustaining memory loops that allow you to 'hold' information in conscious memory for short periods of time, unless rehearsed. Chronic high levels of cortisol due to prolonged traumatic stress such as Post Traumatic Stress Disorder (PTSD) can damage the hippocampus.

Inhibitors *Inhibitors* are molecules that bind to other molecules, including nutrients like zinc that then prevent these 'bound' nutrients from being taken up or properly processed in the body. Phytates are *inhibitors* that bind firmly to zinc & iron and block uptake across the intestinal wall leading to elimination form the body. For example, one of the primary reasons for zinc deficiency is the presence of *Phytates* from cereal products, legumes and nuts. Processed snacks and junk food are particularly rich in these zinc and iron inhibitors.

Inhibitory Neurotransmitter *Inhibitory neurotransmitters* increase the polarity or charge of neuronal membranes making the membranes harder to 'fire', and thus decrease or totally inhibit neuronal activity. Inhibition plays a primary role in higher cortical functions of the frontal lobes like decision-making, and the 'bad' or socially inappropriate behaviour of a drunk is due to loss of frontal lobe inhibition. GABA and Serotonin are two important inhibitory neurotransmitters in the brain.

Iodide *Iodide* is the binary compound of iodine that is an essential co-factor for the production of Thyroxine in the thyroid gland, and thus controls basal metabolism of the body.

When iodide is chronically deficient, the body makes more thyroid tissue trying to compensate for the lack of thyroxine, and this over-production of thyroid tissue that swells the neck is called a goiter.

Kryptopyrrole One component of the haem molecule is porphyrin. The metabolism of haem or haemoglobin releases the porphyrin molecule to be conjugated and excreted in the urine. A faulty gene coding for the enzyme breaking down the haem molecule may produce a different form of porphyrin metabolite originally called Kryptopyrrole, now known to be haemopyrrole. Unfortunately, the kryptopyrrole conjugates with a vitamin B_6 molecule, and the kryptopyrrole-B_6 complex then chelates an atom of zinc, and all three are excreted in the urine.

Kryptopyrroluria (Mauve Factor) The nutritional deficiency disease caused by the faulty gene whose enzyme produces kryptopyrrole instead of normal porphyrin metabolites from the metabolism of haem or haemoglobin. The kryptopyrrole conjugates vitamin B_6 and the kryptopyrrole-B_6 complex then chelates an atom of zinc resulting in a widespread zinc vitamin B_6 deficiency throughout the body and brain. Normally other biochemical compensatory mechanism 'hide' this problem, the reason it is called Krytpo ('hidden' in Greek)-pyrroluria.

However, under stress these compensations 'break down' and mild to severe deficiencies of zinc and vitamin B_6 are created, and the deficiency of these vital nutrients then produces a wide variety of physical, physiological and psychological symptoms. The condition can vary from mild to severe expression of these symptoms including concentration and memory problems, ADHD to schizophrenic-like behaviour.

Limbic System The *Limbic System* is composed of both ancient 'palaeocortex' and various subcortical brain nuclei (islands of grey matter) in the white matter of both the brain, and brainstem. Together these Limbic structures regulate social interactions by generating the social emotions of guilt, shame and blame, and also modulate our brainstem survival emotions like fear and rage in socially appropriate ways. So while you may be very angry at someone over what they did or said, you can control your desire to yell at them or hit them. The function of the Limbic system is largely subconscious, and thus while your know you are angry, often you do not know why!

When your Ego is perceived to be under attack, the Limbic System may activate the 'Fight or Flight' reactions for Ego defense and survival. Once activated, the 'Fight or Flight' reactions initiate a whole cascade of physiological events, and stereotypical 'hardwired' survival reactions, not careful consideration and evaluation of the choices. However, to run these stereotypic reactions uses less than half the energy of thinking.

The *Limbic System* is also linked to muscle tone and tension which is why physical changes occur as emotions change, i.e. your neck tightens as you become stressed or your shoulders slope when you are depressed.

Marginal Nutritional Deficiency A *Marginal Nutritional Deficiency* is a lack of one or more specific nutrients in large enough quantities to handle 'peak' demands. For example, studies have shown that up to 50% of people are zinc deficient yet zinc is an important co-factor & nutrient supporting brain function. So when you have a chronic deficiency of zinc, you often suffer memory and thinking problems.

Also, a deficiency of nutrients, especially vitamins and minerals can significantly reduce enzyme activity. This is a major reason people feel so tired when they are deficient in the B-vitamins that are required to generate ATP, the energy currency of the body.

Also see *Absolute Nutritional Deficiency.*

Neurotransmitters (NT) *Neurotransmitters* are molecules that act as chemical messengers carrying nerve impulses (messages) from one nerve cell to another across the synapse. Clearly, without sufficient NTs, the signalling in the brain is disrupted causing a loss of Brain Integration resulting in the loss of specific mental functions.

Neural Signalling *Neural Signalling* is the communication between the various processing centres within the brain that uses the NT signalling molecules to maintain integrated brain function underlying the capacity for effective memory and thinking. *Neural Signalling* in the brain relies upon stable and flexible neuronal membranes and this requires sufficient levels of the Omega-3 fatty acids, DHA and EPA.

Non-Essential *Non-Essential nutrients* are those that can be generated from other nutrients that are abundant in the normal diet, and therefore do not necessarily have to come directly in the diet.

For example, the important inhibitory NT, GABA, can be made from the molecule glutamate in a single enzymatic step, and glutamate can also be made from the precursor amino acid glutamine also in one step. So GABA is seldom deficient as glutamine is a very common amino acid in many foods.

Nootropic *Nootropic* substances are a new class of drugs, herbs or other substances that enhance cognitive functions with little or no side effects or toxicity. It is derived from the Greek words *noos*, meaning *mind*, and *tropein*, meaning *toward*.

Nootropic Herbs *Nootropic herbs* are herbs that enhance or improve learning and memory. *Gingko biloba* improves cerebral circulation, mental alertness, and overall brain function. *Barcopa monniera* (also called *Brahmi*) facilitates learning, improves memory and is a powerful antioxidant. *Ginseng* helps the body and brain adapt to physiological and psychological stressors, having the ability to evoke a stable state without initiating Fight or Flight reactions, and is thus called an *adaptogen*.

Noradrenalin *Noradrenalin* is an excitatory neurotransmitter involved in alertness, concentration, aggression and motivation among other behaviours. Like Dopamine, it is an important neurotransmitter that controls how 'good' we feel, and how well we handle stress. It helps modulate brain function, especially the Fight or Flight Reaction, and it also helps to modulate our moods, gives us energy, increases our stamina and aids memory. Noradrenalin is created in the body from the amino acid precursor tyrosine or phenylalanine, both essential amino acids.

NT See *Neurotransmitters*.

GLOSSARY OF TERMS

Nutriceuticals *Nutriceuticals* are a whole new concept in nutrition, nutritional formulas designed to provide all of the nutrients necessary to optimise specific functions. Like pharmaceuticals (which are drugs developed to treat specific diseases and conditions), nutriceuticals provide the full complement of nutrients necessary to optimise particular functions – for instance, in the case of ThinkingAdvantage, mental performance by maintaining brain integration under stress.

Nutritional Deficiency *Nutritional Deficiencies* can be either a marginal or absolute.
See Marginal and Absolute Nutritional Deficiencies.

Omega-3 *Omega-3* is a group of polyunsaturated fatty acids that play an important role in the body, particularly the brain. Because they are 'unsaturated', they increase the 'flexibility' and thus function of neuronal membranes. Deficiency of Omega-3 fatty acids may result in slower nerve transmission, and loss or reduced efficiency of many integrated functions such as vision, memory and even coordination of movement. EPA and DHA are two of the most important Omega-3 fatty acids in the brain. The levels of Omega-3 fatty acids in the typical Western diet has decreased by 80% in the past 100 years.

Omega-6 *Omega-6* is a group of polyunsaturated fatty acids that are particularly important in the structure of cell membranes. Arachidonic Acid is an important Omega-6 fatty acid that while important in the structure of nerve membranes, when in excess also leads to the production of inflammatory prostaglandins in tissue resulting in the release of Histamine into these tissues, which causes an inflammatory response accompanied by tissue swelling.

The levels of Omega-6 fatty acids in our bodies, especially Arachidonic Acid, have been greatly increased in Western diets by the presence of 'Trans-' fatty acids so common in fast and junk foods, and virtually all margarines because they cannot be used to make Omega-3 fatty acids, indeed one of the reasons for the decrease in Omega-3 fatty acids.

Oxidation A chemical reaction in which an electron is taken from a molecule of the oxidized substance to form either a chemical bond or an ion. Because oxygen is commonly the source of this electron, this process was referred to as an oxidation reaction or simply oxidation.

Phenylalanine *Phenylalanine* is the amino acid precursor for the NT Dopamine. Dopamine is involved in both the reward and movement systems, but can only be produced if there is sufficient ancillary nutrient matrix containing B_6, zinc, iron, magnesium and calcium present.

Phytates *Phytates* are natural inhibitor molecules found in cereal grains, legumes and nuts to prevent enzyme action while the seeds and nuts are awaiting germination (the reason you can store wheat kernels, dried peas and nuts for long periods).

However, because they are big molecules that are not easily absorbed, when present in foods we eat, they bind so firmly to zinc and iron that they prevent these vital nutrients from being absorbed and used by the body.

Precursor A chemical molecule which provides the primary component of the final product by addition of various functional groups (See Functional Groups).

Precursor Molecules *Precursor Molecules* are the building blocks of more complex molecules needed in the body. Precursors are nutrient molecules (amino acids, simple fatty acids or sugars) that are required for the production of more complex molecules via enzymes adding one or more functional groups or re-arranging the functional groups of the precursor molecule. For example; EPA is the precursor converted into DHA; Tryptophan is the precursor converted into Serotonin; and Tyrosine is the precursor converted into Dopamine or Noradrenalin.

GLOSSARY OF TERMS

RDA *Recommended Dietary Allowance.* The officially recommended amount of each nutrient to be taken in on a daily basis to prevent overt nutritional deficiency disease, but it does ***not*** represent the amount needed to be 'healthy'. Stress requires far more nutrients than 'normal' function, and thus people under stress require far more than the RDA to maintain their health. Particularly under stress therapeutic doses of many nutrients are 5 to many times the RDA.

This is the reason why people can be ingesting the RDA of all of the essential nutrients yet still suffer from chronic nutritional deficiencies of one or more nutrients. The RDA also does not indicate which nutrients or which forms of a nutrient need to be taken together (i.e. the nutrient matrix) to enhance the synergy between them.

Receptor A *Receptor* is protein molecule, often with a sugar (glycoprotein) or lipid (lipoprotein) functional group attached to the cell membrane. Receptors may act in one of two primary ways: 1) Receptors for Hormones: Once the hormone 'docks' on the receptor, this activates a chain reaction of molecules inside the cell to produce the chemicals or molecules required to do that hormones job, which may include activating genes on the DNA molecule. For example, when the pituitary hormone Thyroid-Stimulating Hormone (TSH) 'docks' on the TSH receptor in the thyroid gland, it initiates the production and release of Thyroxine into the blood increasing the rate of basal metabolism warming the body.

2) Receptors may activate 'gates' in the cell membranes that then allows absorption of nutrients into the cell. When NMDA 'docks' on NMDA receptors on the nerve membrane, it opens calcium channels letting calcium into the neuron, and this calcium then activates many cellular reactions, including when in high concentrations, apotosis or programmed cell death.

Serotonin *Serotonin* is the calming, inhibitory neurotransmitter involved in vanquishing depression, but also keeping the processing in the brain orderly by synchronising the timing of neural flows. The production of serotonin is dependent upon the precursor amino acid, Tryptophan and adequate levels of an ancillary nutrient matrix containing the vitamins B_6 and C plus the minerals iron, zinc and magnesium. When serotonin levels are too low, there are mood swings and depression often results.

Stimulatory Neurotransmitter A neurotransmitter that decreases the polarity of the post-synaptic membrane, causing the sodium gates to open and the nerve impulse to be passed on to the next neuron. E.g. Actylcholine and Noradrenal are excitatory neurotransmitters.

Stress *Stress* is the result of activation of the Fight or Flight system for both physical and/or psycho-emotional survival. Stress causes the release of the Stress Hormones, Adrenalin and Cortisol and activates the Sympathetic Nervous System preparing the body physiologically to Fight or Flight by increasing heart rate and blood pressure, releasing glucose into the blood stream, increasing the power of muscle contraction, and re-directing the blood flow from our digestive system to our heart, lungs and muscles, and shutting down blood flow to our thinking frontal lobes, as thinking is too slow for survival.

This shift in the control of brain function from conscious cortical control to the stereotypic 'knee-jerk' reactions run by the Limbic-Brainstem Survival Systems results in an overt loss of Brain Integration further blocking our frontal lobe thinking, problem-solving and decision-making abilities.

To over-ride these automatic survival reactions and remain in conscious cortical control requires considerable energy to suppress these survival emotions, greatly reducing nutrient levels in the brain, especially the frontal cortex. If there are any marginal nutritional deficiencies, the frontal lobes simply run out of enough 'energy', and shut-down leaving the reactive Fight or Flight System in full control.

There are varying types of stress: Psychological, Mental, Emotional, Physical and Physiological, any one of which can fire the Fight or Flight reaction.

Stress Hormones The *Stress Hormones Cortisol and Adrenalin* are released by the adrenal glands during states of stress and negatively affect our physiology (increasing our heart rate, blood pressure, etc.), interfering with mental performance by inhibiting frontal lobe thinking and blocking memory.

Chronic high levels of stress hormones are strongly associated with illnesses such as gastritis-ulcers, colitis, high blood pressure, stroke and coronary heart disease. High levels of Cortisol inhibit the immune system, inhibit memory formation, and if on-going can actually damage the Hippocampus, the shor-term memory centre of the brain.

Synergy *Synergy* is the working together of nutrients for greater effect, the sum being more than its parts. This is perhaps the most important factor in maintaining optimum mental function as it is the synergy between the different nutrients, with one nutrient supporting the function of another that allows brain function to be fully optimized. These positive re-enforcing interactions between different nutrients work to maintain brain integration and create optimal mental performance.

This nutritional synergy is commonly overlooked in most current studies that tend to consider or test only one or a few nutrients at a time. In addition, because nutrients work together synergistically, smaller amounts of individual nutrients are often required if they are taken together as a single matrix than if taken separately.

Transporters *Transporter* molecules are needed to get many of the bigger nutrients like zinc and iron ions into the cell as they cannot easily diffuse through the cell membranes like the smaller ions or molecules, e.g. sodium, potassium and chloride ions or carbon dioxide molecules, respectively.

Thus, Transporter molecules play a vital role in the *Absorption* of zinc and other micro-nutrients from the gut into the blood. They also control the transfer of nutrients from the blood into the cells, a process called *Assimilation*.

The great majority of transporter molecules are proteins, and thus coded for by your genes. Since there are different Alleles for each protein due to mutation, there are faster and more efficient Transporters and slower, less efficient Transporters. People who inherit Alleles coding for slower Transporters may not be able to get enough nutrients into their blood from their food or into their cells from the blood even when eating a normal healthy diet. For these people, vitamin and mineral supplementation provide a means to compensate for their slower Transporters and re-establish normal function.

Tyrosine   Tyrosine is the amino acid precursor for the neurotransmitters Dopamine and Noradrenalin, and like phenylalanine, it can only be converted to these important NTs if there is sufficient ancillary nutrient matrix containing B_6, zinc, iron, magnesium and calcium present.

Zinc   Zinc is an important micro-mineral that is involved in over 300 enzymes in the body and plays many vital roles in both body and brain function. For example, short-term and working memories and executive functions of the frontal cortex are highly dependent upon adequate levels of zinc, as is the immune system, liver detoxification system and the production of digestive enzymes.

Zinc comes in many forms, and the form determines largely where in the body it will be utilised, and how effectively it will be utilised:

Zinc sulphate – common, cheap to make and is poorly utilised by the body, especially in the brain with most rapidly of the zinc being excreted in the urine.

Zinc gluconate – a chelated form of zinc that is more rapidly assimilated into the brain and supports mental function and memory.

Zinc arginate – a chelated form of zinc that strongly supports immune system, liver detoxification function and digestive enzyme function

Zinc oratate – a chelated form of zinc that carries zinc into the mitochondria, the 'powerhouses' of the cell, to assist as a co-factor in enzymes involved in the production of cellular energy in the form of ATP.

Zinc citrate – a chelated form of zinc that appears to facilitate digestive and liver function

Zinc picolinate – a chelated form of zinc that is not found in appreciable amounts in nature, yet is common in many supplements, and may not be well transported in the body.

ANNOTATED LIST OF SUGGESTED READING

Section 1: Suggested Readings on Food, what you should eat and diets to make this possible

1. Lawrence, F. **Not on the Label. What Really Goes into the Food on your Plate.** Penguin Books, London, 2004. (*Excellent, if not somewhat depressing, presentation how the food you eats gets to your table and what has been done to it along the way!*)
2. Light, L. **What to Eat: The 10 Things you really need to Know to Eat Well.** McGraw Hill Publishers, 2005. (*Sound principles of what you should eat and why this is so, with suggested dietary plans by former USDA nutritionist and architect of the original version of the Food Guide Pyramid, not the current guide published by the USDA, a bastardised version altered by the pressure of the food lobby.*)
3. Nestle, M. **What to Eat: An aisle-by-aisle guide to savvy food choices and good eating.** North Point Press, 2006. (*As the title suggests, it provides a reference guide of what to eat from each aisle of your local supermarket, but also lucidly discusses 'Big Food' and how the best decisions for the food corporations are often the worst ones for our health because they must continually increase profits by getting people to eat more of the cheap, processed foods they make!*)
4. Noakes, M. & Clifton, P. **The CSIRO Total Wellbeing Diet.** Penguin Boks, 2005, 208pp. *The CSIRO is the premier scientific*

research organisation in Australia, and CSIRO scientists developed *The Total Wellbeing Diet*, as an easy-to-use, scientifically tested, and nutritionally balanced eating plan. The Total Wellbeing Diet is founded on clinical studies that proved a low-fat, high-protein eating plan allowed participants to eat less while still feeling satifisfied and getting the necessary nutrition, not to mention losing weight in most cases and the weight stayed off. There are also The Total Wellbeing Diet Cookbooks to assist you in your diet change by providing meals and cooking suggestions to get you started. This booked knocked the Da Vinci Code off the number one spot in Australia.

5. Mercola, J., Vaszily, B. & Bentley, N.L. **Dr. Mercola's Total Health Program: The Proven Plan to Prevent Disease and Premature Aging, Optimize Weight and Live Longer.** Mercola.com, 2003–2005. *Dr. Mercola's program comes from decades of his experience in clinical practice, decades spent learning from extensive research, conferring with his professional colleagues, and most of all, successfully treating many thousands of patients. His program is based upon Metabolic Typing, and is complete with recipes and cooking suggestions and ways of altering your food choices to assist you in changing your diet. See: **www.mercola.com** for more information.*

6. *Note: The CSIRO Total Wellbeing Diet and Dr. Mercola's Total Health Program take quite different approaches to a 'healthy diet' with the CSIRO program emphasising cereal grains, balanced with fats and proteins, while Dr. Mercola's program is largely a 'no grains' diet, but different people with different genetics may benefit from different types of diets, and I suggest you see which works best for you.*

Section 2: Suggested Reading on books discussing the issue of Fast & Junk Foods & Why they exist

7. Crister, G. **Fat Land: How Americans Became the Fattest People in the World.** Houghton Mifflin; New York, 2003.

(*Highlights many of the points made in Chapter 2, and clearly exposes the problem – simply too much junk food eaten by too many Americans coupled with too little exercise!*)
8. Brownell, K. & Horgen, K.B. **Food Fight. The Inside Story of the Food Industry, America's Obesity Crisis and What we can do about It.** McGraw Hill Publishers, 2004, pp. 356. (*It is practically criminal the way the food industry has manipulated the government watchdogs like the USDA and FDA who are supposed to be guarding the health of its citizens, certainly an issue to become aware of, as it is the only way this will change.*)
9. Schlosser, E. & Wilson, C. **Chew on This: Everything you don't want to know about fast food.** Houghton Mifflin/Puffin Books, 2006. (*The authors focus on the end result of the profit-above-all-else economics in the fast food and meat-packing industries, providing many examples of why this industry is so powerful and can block attempts to regulate or control it.*)
10. Pollan, M. **The Omnivore's Dilemma: A natural history of four meals.** Penguin Press, 2006. (*A philosophical romp through different meals: fast foods, sustainable farmed and hunted-and-gathered, with a focus on how fabulously expensive ecologically fast foods really are using the overproduction of corn in the U.S. as an example and how clever marketers find a way to induce people to eat this excess. Much of fast food is basically corn in disguise, and soda is little more than high-fructose corn syrup and water.*)

Section 3: Suggested Reading of accessible Nutritional References & Guides to Nutrients

11. Holford, P. **The New Optimum Nutritional Bible.** Basic Health Publications, 2005. (*A comprehensive guide to vitamins and minerals, and the role they play in your health and function.*)

12. Mindell, E.G. **Earl Mindell's Vitamin Bible**. Warner Books, September 2004. (*A complete list of all the nutrients and the role each vitamin and mineral plays in health and well-being.*)
13. Koch, M.U. **Laugh With Health, Renaissance of New Age Australia**. First Ed., August 1981, 19th (Colour) Ed. November 2004. (*Details of nutritional content, glycemic index and discusses the nutritional value of hundreds of foods. Laid out in easy-to-follow charts and addresses numerous commonly asked questions about nutrition.*)
14. Reavley, N. **Vitamins etc**. Bookman Press, Melbourne, 1999. (*A bit older, but excellent guide to nutrients and the role they play in your health.*)
15. Udo, E. **Choosing the Right Fats: For Vibrant Health, Weight loss, Energy, Vitality**. Alive Books, 2002. (*A more detailed presentation of the role fatty acids play in your health, with a bit more detailed presentation of fatty acid metabolism, and how fatty acids are important in the function of both your brain and body.*)

Section 4: Suggested reading on the Effects of Nutrients on Brain Function & Perfromance.

16. Holford, P. **Optimum Nutrition for the Mind**. Basic Health Publications, 2004. (*Excellent clearly written presentation highlighting the major nutrients needed for healthy brain function, with the role each plays in brain function.*)
17. Holford, P. and Colson, D. **Optimum Nutrition for your Child's Mind: Maximising your Child's Potential**. Piatkus, 2006. (*An excellent up to date, well-written guide to how to feed your child to assist their brain's development so that they may reach their true potential.*)
18. Schmidt, M.M. **Brain-Building Nutrition: The Healing Power of Fats and Oils**. 2nd Ed., North Atlantic Books, June,

2001. (*A fairly accessible discussion of the role fatty acids and oils play if helping you achieve effective brain function.*)
19. Stordy, B.J. and Nicoll, M.J. **The LCP Solution: The Remarkable Nutritional Treatment for ADHD, Dyslexia and Dyspraxia**. Ballatine Publishing Group, London, 2000. (*A good summary of the research that has shown how fatty acid deficiencies, especially of Omega-3 fatty acids, affect learning and coordinated movement.*)
20. Khalsa, D.S. **Brain Longevity**. Century Books, London, 1997. (*A good discussion of how to support your brain function, particulalrly as you age, with nutrition, including herbs.*)
21. Carper, J. **Your Miracle Brain**. HarperCollins Publishers, N.Y. 2000. (*A readable book about the brain and brain function for the general reader.*)
22. Sahelian, R. **Mind Boosting Secrets. Natural Supplements that Enhance your Mind, Memory and Mood**. Bottom Line Books, St. Martin's Press, 2005. (*A good summary of the nutrients that have been found to enhance brain power, although a bit of an infomericial for the products they sell.*)

Section 5: Suggested Reading on Stress, Emotions, and Brain Function & Brain Integration.

23. Seyles, H. **The Stress of Life**. Revised Edition. McGraw-Hill Paperbacks, 1978. (*A great read by the person who discovered the General Adaptation Syndrome, about what stress is and how it affects your health.*)
24. LeDoux, J. **The Emotional Brain. The Mysterious Underpinnings of Emotional Life**. Simon & Schuster, New York, N.Y., 1996. (*Slightly more technical, but yet highly readable story of the discovery of the role the Amygdala and Survival System play in our emotional life, and how this largely controls our thinking.*)

25. Krebs, Dr. C.T. & Brown, J. **A Revolutionary Way of Thinking. From a Near Fatal Accident to a New Science of Healing.** Hill of Content Publishing, Melbourne, Australia, 1998. (*A fanscinating personal story, and excellent description of Brain Integration and its role in learning and learning problems, but with a somewhat more technical description of the brain and brain function, including a description of the new science of Kinesiology and how the Acupuncture System interacts with the other Energy Systems of the body.*)
26. Goldberg, E. **The Executive Brain. Frontal Lobes and the Civilized Mind.** Oxford University Press, New York, 2001. (*A wonderful and highly readable story of the role the Executive Functions play in your life.*)
27. Lipton, B. **The Biology of Belief. Unleashing the Power of Consciousness, Matter, & Miracles.** Mountain of Love/Elite Books, Santa Rosa, California, 2005. (*A highly recommended, very accesible presentation of how both our thoughts and environment influence our genetic expression – the new field of Epigenetics.*)

REFERENCES AND CHAPTER END NOTES

(Authors notes & comments are in italics)

Chapter 1: Nutritional Deficiencies: What are they and Why do they Exist?

1. *The brain is indeed the most energy demanding part of our body, and much of this energy goes into maintaining neurotransmitter levels:* Behar, K. L. and Rothman, D.L. In Vivo Nuclear Magnetic Resonance Studies of Glutamate-Á-Aminobutyric Acid-Glutamine Cycling in Rodent and Human Cortex: the Central Role of Glutamine. J. Nutrion 131:2498S–2504S, 2001; Shulman, R.G. Functional Imaging Studies: Linking Mind and Basic Neuroscience. Am. J. Psychiatry 158:11–20, 2001.

2. *As so often happens in science, major discoveries are often made by serendipity:* Christiaan Eijkman, Shared the Nobel Prize in Physiology and Medicine in 1929 with Frederick Hopkins. He was Director of the Geneeskundig Laboratorium (Medical Laboratory), Batavia, Java in the Dutch East Indies (now Jakarta, Indonesia). He discovered that the real cause of beriberi was the deficiency of some vital substance in the staple food of natives, which is located in the so-called 'silver skin' (pericarpium) of the rice. This discovery led to the concept of vitamins. Source: **Nobel Lectures, Physiology or Medicine 1922–1941.** Elsevier Publishing Company, Amsterdam, 1965.

3. *For more information on the effects of Vitamin & Mineral deficiency, you are referred to Appendices 1 & 2:* Muscle weakness from B_1 deficiency: Extreme deficiency of Thiamine (B_1) leads to build up of pyruvic and lactic acids and insufficient production of ATP for muscle and nerve function leading to *beriberi* – partial paralysis of the smooth muscle of the GI tract and skeletal muscle paralysis; and atrophy of the limbs. Tortora, G.J. & Grabowski, S.R. **Principles of Anatomy and Physiology.** 9th ed., John Wiley & Sons, Inc., Brisbane 2000, p. 906.

4. *Sugar in itself is not a bad food, it's just the amounts in your diet & its purity that are problematic:* Cox, C. Killing ourselves with the drug . . . Sugar. The Pioneer, January 7, 2004, Phoenix, AZ; Schauss, A.G. **Diet, Crime and Delinquency.** Parker House Enterprises, 1981; Reuben, D. **Everything You Always Wanted to Know About Nutrition.** Simon and Schuster, New York 1978; *Sugar metabolism to fats also uses vital nutrients from healthy cells:* Reuben, D. ibid, p. 123.

5. *Sugar makes up a lot of kilojoules of many children's diets, which is not surprising when you consider that Coco Puffs are 56% sugar, and even 'low sugar, healthy' breakfast cereals weigh in at 20 to 30% sugar.*

6. *Effects of sugar on children's metabolism and behaviour:* Tamborlane, W.V. Professor of Pediatrics, Yale School of Medicine, and Jones, T.M., visiting scientist from Australia. Children respond more in blood sugar response than adults. Reported in the New York Times, 1990; Conners, C.K. Not sugar alone but sugar and imbalanced diet low in protein. Reported in Med. Tribune, Jan 9, 1985; Wells, K.C. Laboratory of Behavioural Medicine, Children's Hospital, Washington DC. 1985; Goldman et al., Behavioural effects of sucrose on preschool children. J. Abnormal Child Psychology 14(4):565–577, 1986; Denton, D. et al., The impact of the supply of glucose

to the brain on mood and memory. Nutrition Review 59(1, Pt. 2):S20–21, 2001.
7. *Western diet deficient in important nutrients*: Lonsdale, D. & Shamberger, R. Red Cell transketolase as an indicator of nutritional deficiency. Am. J. Clin. Nutr. 22(2):205–211, 1980; Schauss, A.G. Nutrition and Behaviour. J. Applied Nutrition 35(1):30–35, 1983; Sali, A. Dietary fats in health and disease. Australian Family Physician 19:315–320, 1990; Davies, S.A., Stewart, A. **Nutritional Medicine**. Pan Books: London, pp. xxiii–xxiv, 1987; Coghlan, A. Simple food guide goes to heart of the matter. New Scientist, 144 (1951):10, 1994; Anonymous. Putting Fact Foods to the Test. Choice (April), p. 7–9, 1994; Werbach, M. Recent Advances in the Prevention and Treatment of Diseases with Nutrients. Lecture presented in Melbourne, Australia, October 24th 1995.
8. *Iron & Zinc deficiency widespread and associated with Attention Deficit Hyperactivity Disorder (ADHD):* Akhondzadeh, S. Mohammandi, M-R. & Khademi, M. Zinc sulfate as an adjunct to methylphenidate for treating attention deficit hyperactivity disorder in children: A double blind trial. BMC Psychiatry 4(9): 2004 – URL: **www.biomedcentrl.com/** 1471–244X/4/9; Konofal, E., Lecendreux, M., Arnulf, I & Mouren, M-C. Iron Deficiency in Children with Attention-Deficit/Hyperactivity Disorder. Archives of Pediatrics & Adolescent Medicine, Vol. 158(12): 1113–1115, 2004, ncbi.nlm.nih.gov.
9. For information on the movie, Supersize Me, you are referred to the website: **www.supersizeme.com**.

Chapter 2: Reasons for Nutritional Deficiencies
1. *Decrease in mineral content of soils:* 1936 USDA Report to Senate, Senate Document # 264, 74th Congress, 2nd Session. This trend in the decrease of minerals in soils worldwide was highlighted in the June 1992 earth Summit Final Report.

2. *Effect of current farm practices on produce:* Department of Environment, Food and Rural Affairs: Food and farming: A Sustainable Future, report by the UK Policy Commission on the Future of Farming and Food, chaired by Sir Don Curry, January 2002, available from **http://www.cabinet-office.gov.uk**.; *and for a recent extensive discussion of current farming and food processing practices and their effect on food quality, see*: Lawrence, F. **Not on the Label. What Really Goes into the Food on your Plate**. Penguin Books, London, pp. 272, 2004.

3. *Modified-Atmosphere packaging (MAP) can now increase shelf-life, but at what cost to food quality and health*: Xiong, Li, Modified Atmosphere Packaging, Fact Book (Department of Food Science, Pennsylvania State University, 1999: Smyth, A.B. et al., Modified atmosphere packaged cut iceberg lettuce: effect of temperature and CO_2 partial pressure on respiration and quality. J. Agricultural Food Chemistry, 46, 4556–4562, 1998; Serafini et al. Effect of acute ingestion of fresh and stored lettuce on plasma total antioxidant capacity and antioxidant levels in human subjects. British J. of Nutrition, 88: 615–623, 2002. *For cold store and the handing of food you buy at the supermarkets, see:* Lawrence, F. **Not on the Label. What Really Goes into the Food on your Plate**. Penguin Books, London, pp. 272, 2004.

4. *The decrease in nutrient concentration of basic foods is both pervasive and shocking*: Mayer, A. Historical changes in the mineral content of fruits and vegetables: A cause for concern? British Food Journal 99/6, S. 207–211, 1997; *Thomas, D.* A case for the need for nutritional supplementation. Cranial View, May 2000 (vgl.:http://www.craniogroup.com); Leake, J. 'It's not the fruit it used to be.' in: The Sunday Times, 8.2.2004, London (**http://timesonline.co.uk/article/0,,8126-995115,00.html.**); Bundesministerium für Verbraucherschutz, Ernährung und Landwirtschaft (Hrsg): *Ernährungs – und agrarpolitischer Bericht*

der Bundesrierung, Bonn, 2003. (vgl.: **www.verbraucherminis terium.de**); Lammery, W. Mangel im Überfluß. In: Newsletter Heilprakitiker vom 18 Juli 2004, #639. Private analyses by GmbH, Handels- und Umweltsschutzlabor, Dr. Kaiser & Dr. Waldman GmbH, 27761, Hamburg, Germany, May 2004.

5. *What effect reduced nutrients in food is having on dietary nutritional deficiencies is open to question, but it has to play a significant role in why western diets lead to marginal nutritional deficiencies*: Food Standards Agency (FSA), National Diet and Nutrition Survey, UK. (**http://www.timesonline.co.uk/newspaper/0,, 176-993250,00.html**)
Organic foods contain less pesticides and food additives that have been linked to increased cancer and allergies: Baker, B.P., Benbrook, C.M., Groth, E. and Benbrook, K.L. Pesticide residues in conventional, IPM-grown and organic foods: Insights from three U.S. data sets. Food Additives and Contaminants, 19(5): 427–446, 2002; Charlier G., et al. Breast cancer and serum organochlorine residues. Occupational and Environmental Health Perspectives, 60 (50: 348–351, 2003; Schreinemachers, D.M. Use of agricultural pesticides and prostrate cancer risks in the agricultural health study cohort. Am. J. Epidemeology, 157,(9): 800–814, 2000; Curl, C.L., Fenske, R.A., and Elgethun, K. Organophosphorus pesticide exposure of urban and suburban pre-school children with organic and conventional diets. Environmental Health Perspectives, 111 (3):377–382, 2003; Stapleton, C. Toxic elements found in infants' cord blood. Palm Beach Post, July 15,2005, **http://www.palmbeachpost.com/news/content/ nation/epaper/2005/07/14/m1a_peststudy_0714.html**

6. *There is a **real** difference in the nutritional quality and levels of contaminats between Organic grown versus conventional grown Factory Farm foods:* Davison, J. **Cancer Winner**. Pierce City, MO. Pacific Press, 1977; McCance and Widdowson, The

composition of foods. MAFF and the Royal Society of Chemistry, 1991; Brandt, K. and Molgaard, J.P. Organic Agriculture: Does it enhance or reduce the nutritional value of food plants. J. Science in Food and Agriculture, 81:924–931, 2001. Carbonaro, M. et al. Modulation of antioxidant compounds in organic versus conventional fruit (Peach, Prunus persica L. and Pear, Pyruscommunis L.). J. Agricultural and Food Chemistry 50'5458–5462, 2002; Asami et al. Comparison of the total phenolic and ascorbic acid content of freeze-dried and air-dried Marionberry, Strawberry and corn using conventional, organic and sustainable agricultural practices. J. Agricultural and Food Chemistry, 51:1237–1241, 2003 & **http://www.mercola. com/ 2003 /mar/22/antioxi dants.htm**; Randerson, J., Microwave cooking zaps nutrients. Reported in J. Sci. Food & Agriculture, Vol. 83:p. 1511, 2003; Surprising Health Benefits of an Organic Diet. **Danish Research Centre for Organic Farming**, reported in **Science Daily** March 29, 2005 & **http://www.mercola.com/ 2005 /apr/13/organic_diet.htm**.

7. *While the GM foods issue is hotly debated, as of 2004 there were only 10 published studies on the health effects of GM food, but clearly as with the Green Revolution rice, there may be unintended results from widespread use of these foods:* Anonymous. Fortifying your food crops. New Scientist Oct. 2003, p. 6; Pryme, I.F. & Lembcke, R. In vivo studies on the possible health consequences of genetically modified plant materials. Nutrition and Health, 17:1–8, 2003; Trials of Rice, In Brief: New Scientist 2 April 2005. Glufosinate ammonium and glyphosate effects and use see: **www.mercola.com/2003/jul/2/gm_crops.htm**; Wijers-Hasegawa, Y. Males more prone to commit crimes but whys elude. The Japan Times: Dec. 7, 2004, **http://www. japantimes.co.jp/cgibin/getarticle. pl5?nn20041207f1.htm.**; The Independent, 22 May 2005.

8. *RDAs were established to prevent nutritional deficiency diseases resulting form Absolute Nutritional Deficiencies, not for optimum health:* Bloomfield, M.M. & Lawrence, J.S. *Chemistry and the Living Organism.* 6th Ed., John Wiley & Sons, N.Y., 1996, p. 475; Werbach, M.R. **Nutritional Influences on Illness. A Source Book of Clinical Research.** Thorsons Publishing Group, London, 1987, p. 466; **Dietary Reference Intakes: Applications in Dietary Assessment.** Institute of Medicine, National Academy Press, Washington, DC, 2001; **Dietary Reference Intakes for Vitamin C, Vitamin E, Selenium and Carotenoids, Institute of Medicine.** National Academy Press, Washington, DC, 2001; **Dietary Reference Intakes for Vitamin A, Vitamin K, Arsenic, Boron, Chromium, Copper, Iodine, Iron, Manganese, Molybenum, Nickel, Silicon, Vanadium and Zinc.** Institute of Medicine, National Academy Press, Washington, DC, 2001.

9. *Values for optimum health and mental function are generally 5 to 50 times the RDA for most nutrients:* Fraga, C.G. et al. Ascorbic acid protects against endogenous oxidative damage in human sperm. Proc Natl Acad Sci 88(24):11003–6, 1991; Bradley, D. Science: An orange a day helps to keep your sperm OK. New Scientist, 14 March, 1992, p. 20–21; Werbach, M.R. **Nutritional Influences on Mental Illness: A Source Book of Clinical Research.** Thorsens Publishers Ltd, London, 2nd ed., Appendix C: Guidelines for Nutritional Supplementation, 1991. p. 466; *Note: In a study of 4,700 patients at risk of developing Alzheimer's Disease (AD) supplementing with Vitamin C & E at doses more than 15 and 55 times the RDA respectively, was associated with a 78 percent lower risk of AD. In the second phase, the percentage dropped to 64, but still indicated a significant level of protection. Supplementing with a typical multivitamin that contained no more than the RDA's had no effect on preventing AD*; Zandi, P.P. Reduced Risk of Alzheimer Disease in Users

of Antioxidant Vitamin Supplements. The Cache County Study. Archives of Neurology, Vol. 61(1): 82–88. 2004, archneur.ama-assn.org; 'Vitamins C, E Linked to Strong Reduction in Alzheimer's Risk' **www.NutraIngredients.com**, 1/20/04, nutraingredients.com.

10. *Bioflavonoids enhance the absorption of Vitamin C and protect it from oxidation: Bioflavonoids were discovered by the Nobel Prize-winning chemist, Albert Szent-Gyorgi, who labeled them 'Vitamin P', as he found they appeared essential for many bodily functions. The common bioflavonoids used to facilitate Vitamin C uptake are citrin, hersperidine, rutin, flavones & flavonals:* Reavley, N. **Vitamins etc.**, Bookman Press, Melb., 1999, p. 364.

11. *The form of the nutrient is important:* Presence of picolinic acid may negatively affect your Zinc status; Seal, C.J. & Heaton, F.W. Effect of dietary picolinic acid on the metabolism of exogenous and endogenous zinc in the rat. J. Nutr. 115:986–993. 1985; *While the synthetic vitamin C molecule is chemically identical to natural forms, the difference arises in other nutrients that accompany the C, especially bioflavonoids, which make the uptake of C more effective; Zinc, Iron and several other larger metal ions require transporter molecules to be taken up into the blood, and into the cells, and zinc has at least 6 different transporter molecules:* Burdette, S.C. & Lippard, S. J. Meeting of the Minds: Metalloneurochemistry. Proc. Natl. Acad. Sci. 100(7):3605–3610.2003, **www.pnas.org/cgi/doi/10.1073/pnas.0637711100**; *Other Studies show that type of magnesium or calcium used is important with more inorganic forms of minerals such as Magnesium and Calcium oxide or Calcium carbonate less well absorbed than chelated forms such as citrates or gluconates*; 'Dietary Magnesium Intake in Relation to Plasma Insulin Levels and Risk of Type 2 Diabetes in Women' Diabetes Care, Vol. 27, No. 1, January 2004, ncbi.nlm.nih.gov.;

'Magnesium May Reduce Risk of Diabetes' Karen Collins, R.D., MSNBC, 5/7/04, www.msnbc.com.
12. *Low Fibre & Fat in Fast Foods*: Sali, A. Proposals for Nutritional Guidelines for Health Education in Britain, prepared for the National Advisory Committee on Nutrition Education (NACNE). Health Education Council, London, Sept. 1983; Davies, S.A., Stewart, A. **Nutritional Medicine.** Pan Books: London, pp. xxiii–xxiv, 1987; Coghlan, A. Simple food guide goes to heart of the matter. New Scientist, 144 (1991):10, 1994; Putting Fact Foods to the Test. Choice (April), pp. 7–9, 1994.
13. *Americans eating little vegetables:* Munz, K. et al., Food intakes of US children compared with recommendations. Pediatrics 100:323–329, 1997; Dennison, B.A. et al., Fruit and Vegetable intake in young children. J. Am. College Nutrition 17(4):371–378, 1998; USA Today August 23, 2004; Light, L. **What to Eat: The 10 Things you really need to Know to Eat Well.** McGraw Hill Publishers, 2005, p. 268.
14. *The Top 25. The biggest selling items in Australian Supermarkets.* Time, June 22, p. 55, 1992.
15. *Junk and Fast Foods now make up a significant percent of many people's diet:* Mercola, J. & Droege, R. Why is Junk Food so Tempting, and How to be Beat your Temptation. 3-17-04, **www.mercola.com/2004/ mar/17/junk_food.htm**; Junk food diet gives one youth scurvy 2-6ä-03, **www.mercola.com/2003/feb/8/junk_food_diet.** htm; Brownell, K. & Horgen, K.B. **Food Fight. The Inside Story of the Food Industry, America's Obesity Crisis and What we can do about It.** McGraw Hill Publishers, 2004, p. 356; Crister, G. **Fat Land: How Americans Became the Fattest People in the World.** Houghton Mifflin; New York, 2003; Junk Food One Third of Diet. Yahoo News, June 2, 2004; Light, L. **What to Eat: The 10 Things you really need to Know to Eat Well.** McGraw Hill Publishers, 2005, pp. 268.

16. *Billions spent on Junk Food Advertising:* http://www.mercola.com/2005/jan/29/ junk_food_ads.htm; Four Ways Junk Food Marketing Targets Kids. 11-26-03. www.mercola.com/2003/nov/26/junk_food_marketing. htm; Wilson, C. Food Kills. Comment & Analysis. New Scientist, 26 Nov. 2003, p.16; *McDonalds and Coca-Cola spend US$1.7 billion annually on advertising:* Webb, J. et al. Hungry for Change. Opinion Futures – Food & Farming. New Scientist 17 July, 2004, pp. 42–45; J. American Dietetics Assoc. 101(1): 42 46, 2001.

17. *Energy density of Fast Food may be the primary cause of obesity in developed countries:* U.S. Junk Food Intake Worsening. 5-08-02, www.mercola.com/fcgi/pf/2002/may/8/junk_food.htm; Coghlan, A. Fast Foods Trick the Body. New Scientist, 25 Oct. 2003, p. 10; Prentice, A. & Jebb, S. Obesity Reviews. Vol. 4:187, 2003; Tanner, L. U.S. Teens Fatter than their Peers in Industrialized Nations. 1-06-04: www.aolsvc.news.aol.com/news/article.adp?id= 20040105-165409990001; Bray, G.A., Nielsen, S.J. & Popkin, B.M. Consumption of High-Fructose Corn Syrup in Beverages May Play a Role in the Epidemic of Obesity American J. Clinical Nutr. 79(4):537–543, 2004. Gross, L.S., Li, L., Ford, E.S. & Lui, S. Increased consumption of refined carbohydrates and the epidemic of type-2 diabetes in the United States: an ecological assessment. American J. Clinical Ntr. 79(5):774–779, 2004.

18. *Information on genetics and Alleles can be found in any standard Anatomy & Physiology textbook:* For example, Tortora, G.J. & Grabowski, S.R. **Principles of Anatomy and Physiology**, 9th ed., John Wiley & Sons, Inc., Brisbane 2000; Rhoades, R. & Planzer, R. **Human Physiology**. 2nd ed., Saunders College Publishing, New York, 1992. *However, as the new science of Epigenetics is showing, genes are not destiny, but rather the environment controls their expression, and your nutritional status is a major factor in gene expression. For a readable discussion of*

epigenetics, see: Lipton, B. **The Biology of Belief. Unleashing the power of consciousness, matter, & miracles.** Mountain of Love/Elite Books, Santa Rosa, California, 2005, p. 224.

Chapter 3: Nutrition and How it Works

1. *This chapter presents the general concepts of nutrition gathered from many different sources, but not before published in one concise presentation at level for the 'average' person to comprehend. The Key Concepts are given in Concept Boxes for you to clearly denote the main principles of nutrition.*
2. *Several sources that are excellent for accessing nutritional information are the following:* Dr. Mecola's website: **www.mercola.com**; the health web newsletter published by Health Science Instute: **HSIResearch@healthiernews.com**; Reavley, N. **Vitamins etc.** Bookman Press, Melbourne, 1999, pp. 792; Kirschmann, G.J. **Nutrition Alamanac.** 4th Edition, McGraw-Hill, New York 1996; Mindell, E.G. **Earl Mindell's Vitamin Bible.** Warner Books, September 2004; Holford, P. Optimum Nutrition for the Mind. Basic Health Publications, 2004; Holford, P. The New Optimum Nutritional Bible. Basic Health Publications, 2005.
3. *The two figures at the end of this chapter summarise nutrients, what they are, what they do in the body and where they come from. Appendices 1 & 2 at the end of the book cover this information in more detail nutrient by nutrient.*

Chapter 4: The Zinc Connection

1. *Zinc deficiency in human populations:* Pfeiffer, C.C. **Zinc and Other Micro Nutrients.** Keats Publishing, Connecticut, Chapter 1–18, 1978; Pfeiffer, C.C. & Bravermann, E.R. Zinc, the Brain and Behaviour. Biol. Psychiatry, 17(4): 1982; Luecke, R.W. Domestic animals in the elucidation of zinc's role in nutrition. Fed. Proc. 43:2823– 2828, 1984; Sandstead, H.H. Is zinc deficiency a public health problem? Nutrition 11:

87–92, 1995; Golub, M., et al. Developmental zinc deficiency and behaviour. J. Nutri.125:2263S–2271S, 1995; Sandstead, H.H. Causes of iron and zinc Deficiencies and their Effects on the Brain. J. Nutri. 130:347S0349S, 2000; Tapiero, H. & Tew, K. Trace elements in human physiology and pathology: zinc and metallothioneins. Biomed & Pharm, 57:399–401, 2003; Fraga, C. Relevance, essentiality and toxicity of trace elements in human health. Mol. Aspects Med. 26:235–244, 2005.

2. *Phytates and their role in reducing the bioavailability of zinc:* Flangan, P.R. A model to produce pure zinc deficiency in rats and its use to demonstrate that dietary phytate increases the excretion of endogenous zinc. L. Nutrition 114: 493–502, 1984; Torres, M. Rodriguez, A.R. & Saura-Calixto, F. Effects of dietary fibre and phytic acid on mineral bioavailability. Crit. Rev. Food Sci. Nutri. 30:1–22, 1991; Williams, A.W. & Erdman, J.W. Food processing: Nutrition, safety, and quantity balances. In Shils, M.E., Olsen, J.A., Shike, M. & Ross, A.C. **Modern Nutrition in Health and Disease.** 9th ed. Baltimore: Williams & Wilkins. Pp. 1813–1821, 1999; Lonnerdal, B. Dietary factors influencing zinc absorption. J. Nutr.,130:1378S–1383S, 2000; **Dietary Reference Intakes for Vitamin A, Vitamin K, Arsenic, Boron, Chromium, Copper, Iodine, Iron, Manganese, Molybenum, Nickel, Silicon, Vanadium and Zinc.** Institute of Medicine, National Academy Press, Washington, D.C, 2001, pp. 457–458.

3. *Zinc Transporter molecules and their role in zinc dynamics – for a review see:* Soong, M.K. & Adham, N.F. Evidence for an important role of prostaglandins and F2 in the regulation of zinc transport in the rat. J. Nutrition 109:2152–2159, 1980; Burdette, S.C. & Lippard, S.J. Meeting of the Minds: Metalloneurochemistry. Proc. Nat. Acad. Sci. 100(7):3605–3610, 2003.

4. *For the role of Zinc as an effective antioxidant.* Prasad, A. Bao, B. Kucuk, O. & Sakar, F. Antioxidant effect of zinc in humans. Free Rad Biol & Med. 37(8):1182–1190, 2004. *& for the role of Zinc in supporting immune function see:* Mocchegiani, E., Giacconi, R., Muzzioli, M. & Cipriano, C. Zinc, infections and immunosenescence. Mech Aging & dev. 121:21–35, 2000.
5. *For effects of genetic zinc deficiency resulting from Kryptopyrroluria see*: Irvine, D.G., Bayne, W, Miyashita, H. & Majer, J.R.: Identification of Kryptopyrrole in Human Urine and its Relation to Psychosis. Nature 224(11): 811–813, 1969; Pfeiffer, C.C., & Illev, V. Pyrroluria Urinary Mauve Factor causes double Deficiency of B_6 and Zinc in Schizophrenics. Fed. Proc. 32:276, 1973; Pfeiffer, C.C. et al. Treatment of Pyrroluric Schizophrenia (Mal varia) with large Doses of Pyridoxine and a Dietary Supplement of Zinc. L. Orthomol. Psychiatry 3:292–300, 1974; Ward, J. Relationship between Kryptopyrrole, Zinc and Pyridoxine in Schizophrenics. J. Orthomol. Psychiatry 4(1):25–32, 1975; Pfeiffer, C.C. **Zinc and Other Micro Nutrients.** Keats Publishing, Connecticut, Chapter 1–18, 1978.

    For the binding of Zinc to a Kryptopyrrole-B_6 complex see: Pfeiffer, C.C., & Illev, V. Pyrroluria Urinary Mauve Factor causes double Deficiency of B_6 and Zinc in Schizophrenics. Fed. Proc. 32:276, 1973; Pfeiffer, C.C. **Zinc and Other Micro Nutrients.** Keats Publishing, Connecticut, 1978 & Maines, M. Zinc-Protoporphyrin is a Selective Inhibitor of Heme Oxygenase Activity in the Neonatal Rat. Biochemica & Biophysica Acta 673: 339–350, 1981. *More recently the actual agent of Kryptopyroluria appears to be not Kryptopyrrole, which is a photo-oxidation product produced in the urine, but rather two stereoisomers that are extremely similar in structure to Kryptopyrrole. These hemopyrroles are actually produced as part of a group of porphyrias all resulting from genetic mutations in haem and* *haemoglobin synthesis and metabolism, of which there are at least 8 known forms. For the purposes of simplicity, I have referred to the genetic disorder that causes zinc and B_6 deficiency as Kryptopyrroluria recognising that this is not exactly correct, but the accurate chemistry is far too complex to discuss in the context of this book, see:* Irvine, D.G., Bayne W. et al. Identification of kryptopyrrole in human urine and its relationship to psychosis. Nature, 224:811–813, 1969; Irvine, D.G. Pyrroles in Neuropsychuiatric and Porphryric Disorders: Confirmation of a Metabolite Structure by Synthesis. Life Sciences 23:983–990, 1978; Irvine, D.G. Hydroxy-hemopyrrolennone, not Kryptopyrrole, in the Urine of Schizophrenics and Porphyrics. Clin. Chem. 24(11):2069–70, 1978; Jackson, J.A., Riordan, H.D., Neathery, S. & Riordan, N. Urinary pyrolles in health and disease. J. Orthomol. Med. 1(2):96–98, 1997. (*an excellent review of this issue*); Rauppinen, K. Porphyrias. Lancet 365:241–252, 2005 (*a good review article*).

6. *For laboratories that can provide the Urine Test for Kryptopyrroluria see:* In Australia: ARL-Pathology, 568 St. Kilda Road, Mebourne VIC 3004. Tel. +61 (0)3 9539 5455, Fax: +61 (0)3 9529 7277 website: http://www.arlaus.com.au.html. In the US:The Bio-Center Laboratory: Tel. +1 316 684 7784; Website: http://biocenterlab.org/tests/urine/pyrroles.shtml. In Germany: Sension GmbH, Am Mittleren Moos 48, 86167 Augsburg, tel. 0821-7493173, fax. 0821-7493171 website: www.sension-gmbh.de; *Currently, in a number of other countries there are no laboratories that perform tests for Kryptopyrrole or pyrroles in the urine.*

7. *To obtain an effective nutriceutical specifically designed to balance the nutritional deficiencies created by Kryptopyrroluria the following source is currently available.* The OrthoKrypto nutriceutical can be ordered from: **Sunflower Deutschland e.V., Lothringenstr. 6b, D-45259, Essen, Germany; Tel. no.**

+49 (0)201 788 477, Fax +49 (0)201 467 504; website: www.sunflower-therapie.com; and cost approximately 30 euros for a month's supply. In Australia you can contact: Oborne Health Supplies Ltd, 13 Harker Street, Burwood, Vic. 3215, Australia. Tel. +61 (0)3 8831 3888, Fax. +61 (0)3 8831 3898 website: www.oborne.com.au.

8. *For review of the roles zinc plays in the body see:* Pan, T. & Coleman, J.E. Structure and function of the Zn(II) binding site within the DNA-binding domain of the GAL4 transcription factor. Proc Natl Acad Sci. May; 86(9): 3145–3149, 1989; **Dietary Reference Intakes for Vitamin A, Vitamin K, Arsenic, Boron, Chromium, Copper, Iodine, Iron, Manganese, Molybenum, Nickel, Silicon, Vanadium and Zinc, Institute of Medicine.** National Academy Press, Washington, DC, 2001; Dreosti, I. Zinc and the gene. Mut. Res. 475:161–167, 2001.

9. *Zinc deficiency and behaviour:* Grant, E.C.G. et al. Zinc deficiency in children with dyslexia: concentrations of zinc and other minerals in sweat and hair. British Med. J. 296: 607–609, 1988; Sandyle, R. Zinc deficiency in attention deficit hyperactivity disorder. Intl. J. Neuroscience 52:239–241, 1990; Mitchell, J. Attention deficit disorder and its treatment. ATOMS, J. Aust. Traditional Med. Soc. pp. 15–17. (Summer) 1994/95; Golub, M., et al. Developmental zinc deficiency and behaviour. J. Nutri.125:2263S–2271S, 1995; Toren, P. et al., Zinc deficiency in attention deficit hyperactivity disorder. Biiol. Psychiatry 40:108–1310, 1996; Sandstead, H.H., et al. Effects of repletion with zinc and other micronutrients on neuropsychologic performance and growth of Chinese children. A. J. Clin. Nutr. 68: 470S–475S, 1998; Sandstead, H.H. Causes of iron and zinc Deficiencies and their Effects on the Brain. J. Nutri. 130:347S0349S, 2000; Stanstead, H., Frederickson, C. & Penland, J. History of zinc as related to brain function.

J. Nutr. 130:496S–502S, 2000; Bhatnagar, S. & Taneja, S. Zinc and cognitive development. British J. Nutri. 85, Suppl. 2:S139–S145, 2001; Nick, G.L. Whole Food Nutrition For ADHD – Medicinal Properties In Whole Foods. Townsend Newsletter for Doctors and Patients, Oct. 2003, **http://www.findarticles.com/p/ articles/mi_m0ISWis_243/ai_1099 46530**; Akhondzadeh, S. Mohammandi, M-R. & Khademi, M. Zinc sulfate as an adjunct to methylphenidate for treating attention deficit hyperactivity disorder in children: A double blind trial. BMC Psychiatry 4(9): 2004 – URL: **www.biomedcentrl.com/ 1471-244X/4/9**.

10. *For information on the Sunflower Program you are referred to:* Otto, Dr. G. **How children with learning difficulties master school successfully through the Sunflower Therapy/Die Sunflower Therapie**, Allitera Publishing House/Allitera Verlag., Germany, 2001. 70pp; *and the Sunflower Therapy Websites:* info@sunflowertherapie.com; www.sunflower-therapie.com & www.sunflower-international.com *and for information about the LEAP® Program you are referred to my book:* **A Revolutionary Way of Thinking. From a Near Fatal Accident to a New Science of Healing**. Krebs, Dr. C.T. and Brown, J., Hill of Content Publishing, Melbourne, Australia, 1998, Chapter 9, pp. 246–271 available from **www.equilibrium.com.au** & **www.amazon.com**, and the LEAP® Websites **www.leapbrainintegration.com**.

11. *Type of supplementation affects outcome, whether effective or not effective:* Cavan, K.R. et al. Growth, body composition of periurban Guatemalan children in relation to zinc status: a longitudinal zinc intervention study. Am. J. Clin. Nutr. 57:344–352, 1993; Sazawal et al. Effect of zinc supplementation on observed activity in preschool children in an urban slum population. Pediatrics 98:1132–1137, 1996; Sandstead, H.H., et al. Effects of repletion with zinc and other

micronutrients on neuropsychologic performance and growth of Chinese children. A. J. Clin. Nutr. 68: 470S–475S, 1998; Bhatnagar, S. & Taneja, S. Zinc and cognitive development. British J. Nutri. 85, Suppl. 2:S139–S145, 2001.

12. *The information on which types of zinc were most effective for different body systems were primarily the result of preliminary research I conducted using acupressure formatting for organ systems and direct muscle biofeedback. But these are based upon 15 years of clinical experience and observation of what worked and what did not work in clinical practice.*

13. *Zinc toxicity: One study found gastro-intestinal upset at levels of 50 to 150 mg of elemental zinc/day and some decreased serum HDL levels, while another study found gastro-intestinal upset, but no depression of HDLs at 150 mg of elemental zinc per day.* Freeland-Graves et al., Effect of zinc supplementation on plasma high-density lipoprotein cholesterol and zinc. Am. J. Clin. Nutr. 35:988–992, 1982; Samman, S. & Roberts, D.C.K. The effect of zinc supplements on lipoproteins and copper status. Atheroscerlosis 70:247–252, 1988. *The quote on the general lack of zinc toxicity at even relatively high levels of supplementation, not including the zinc already in the diet, is taken from:* **Dietary Reference Intakes for Vitamin A, Vitamin K, Arsenic, Boron, Chromium, Copper, Iodine, Iron, Manganese, Molybenum, Nickel, Silicon, Vanadium and Zinc.** Institute of Medicine, National Academy Press, Washington, DC, 2001, p. 488.

14. *For more information on Thinkingadvantage Organic Zinc and more information on the effects of Zinc on health, you are referred to:* www.thinkingadvantage.com.

Chapter 5. Introduction to Brain Integration

1. *The concept of Brain Integration and its effects on learning and mental performance are discussed in-depth in my book:* Krebs,

C.T. & Brown, J. **A Revolutionary Way of Thinking. From near Fatal Accident to a New Science of Healing.** Hill of Content Publishing, 1998, chapters 5 to 9. See also LEAP® website: www.leapbrainintegration.com

2. *The dependence of brain function upon the synchronicity and timing of neural flows and time-binding are discussed in the following:* Damasio, A.R. **Descartes' Error, Emotion, Reason and the Human Brain.** G.P. Putman & Sons, New York, NY, pp. 94–96, 1994; Nunez, P.L. **Neocortical Dynamics and Human EEG Rhythms.** Oxford University Press, New York, NY, 1995; Krebs, C.T. & Brown, J. **A Revolutionary Way of Thinking.** *Ibid*, 1998, Chap. 9. *Habib introduces a concept similar to Brain Integration in his 'Temporal Processing Impairment Model' of learning and cognitive problems:* Habib, M. The neurological basis of developmental dyslexia: An Overview and working hypothesis. Brain 123:2373–2399, 2000.

3. *Executive functions of the brain and their role in our higher-level thinking is interestingly discussed in:* Goldberg, E. **The Executive Brain. Frontal Lobes and the Civilized Mind.** Oxford University Press, New York, 2001.

4. *The concept of Stress and the concept of the Limbic-Brainstem Survival Systems are presented in:* Seyles, H. **The Stress of Life.** McGraw-Hill Book Co., New York, 1976; McLean, P.D. **The Truine Brain in Evolution. Role in Palaeocerebral Functions.** Plenum Press, New York, NY, 1990. Chrousos, G.P. et al. Editors, **Stress: Basic Mechanisms and Clinical Implications.** Annals N.Y. Acad. Sci. 771, 1995; and Panksepp, J. **Affective Neuroscience. The Foundations of Human and Animal Emotions.** Oxford University Press, Oxford, 1998.

5. *The role of the Amygdala and the brainstem survival systems in controlling our basic emotions is discussed in:* LeDoux, J.E. Emotion and the amygdala. In: Davis, M. The Role of the Amygdala in Fear and Anxiety. Annual Rev. Neurosci.

15:339–351, 1992; LeDoux, J.E. Emotion, memory and the brain. Sci. Am. 280(6):50–57, 1994; LeDoux, J. **The Emotional Brain. The Mysterious Underpinnings of Emotional Life.** Simon & Schuster, New York, NY, 1996.
6. *The role of the corpus callosum in maintaining integrated brain function is derived from:* Kolb, B., & Whishaw, I.Q. **Fundamentals of Human Neuropsychology.** 3rd ed. W.H. Freeman & Company, New York, NY, pp. 412–501, 199; Guyton, A.C. **Textbook of Medical Physiology.** 8th ed., W.B. Saunders Company, Sydney p. 642, 1991; Kandel, E.R., Schwartz, J.H. & Jessell, T.M. (Eds), **Principles of Neural Science.** 4th ed., McGraw-Hill, NY, 2000, pp. 322–323; and other neurology textbooks, and other neurology textbooks, and my own research and over 20 years of clinical practice.

Chapter 6. Effects of Stress on Mental Performance
1. *The Fight or Flight System was described by Walter Cannon in the 1920's, but continues to be a useful concept with ever deeper neurological understanding.*
2. *Fight or Flight reactions take place out of our consciousness, and are totally Limbic and Brainstem in origin, and elicit both physiological and behavioural stereotypic patterns of reaction. One of the most powerful is the redistribution of blood in the brain, with reduced blood flow in the frontal lobes, shutting down our executive thinking functions in favour of stereotypic reactions activated by survival emotions like 'fear'. For descriptions and definitions of Stress and the Fight or Flight Reaction you are referred to:* Seyles, H. **The Stress of Life.** Revised Edition. McGraw-Hill Paperbacks. 1978; LeDoux, J. **The Emotional Brain. The Mysterious Underpinnings of Emotional Life.** Simon & Schuster, New York, NY, 1996; Kandel, E.R., Schwartz, J.H. & Jessell, T.M. (Eds), **Principles of Neural Science.** 4th ed., McGraw-Hill, NY, 2000.

REFERENCES AND CHAPTER END NOTES

Chapter 7. How Nutrition can Optimize Mental Performance

1. *It should be no surprise that supplementing children who are starving with nutrients critical for brain function improves cognitive functions (ability to think and reason) and memory:* Soewondo, M.H. & Pollitt, E. Effects of iron deficiency on attention and learning processes in preschool children, Bandung, Indonesia Am. J. Clin. Nutr. 50:667–674, 1989; Pollitt, E. et al. Iron deficiency and educational achievement in Thailand. Am. J. Clin. Nutr. 50:687–697, 1989; Seshadri, S. & Gopaldas, T. Impact of iron supplementation on cognitive functions in preschool and school aged children: the Indian experience. Am. J. Clin. Nutr. 50:675–686, 1989; Schoenthasler, S.J. et al. Controlled trial of vitamin-mineral supplementation: Effects on intelligence and performance. Personality & Individ. Differences 12(4):351–352, 1991; Cavan, K.R. et al. Growth, body composition of periurban Guatemalan children in relation to zinc status: a longitudinal zinc intervention study. Am. J. Clin. Nutr. 57:344–352, 1993; Benton, D. et al. The impact of long-term vitamin supplementation on cognitive functioning. Psychopharmacology (Berl.) 117(3): 298–305, 1995; Sazawal et al. Effect of zinc supplementation on observed activity in preschool children in an urban slum population. Pediatrics 98:1132–1137, 1996; Snowden, W. Evidence from analysis of 2000 errors and omissions made in IQ tests by a small sample of schoolchildren, undergoing vitamin and mineral supplementation, that speed of processing is an important factor in IQ performance. Personality & Individ. Differences. 22(2):131–134, 1997; Sandstead, H.H., et al. Effects of repletion with zinc and other micronutrients on neuropsychologic performance and growth of Chinese children. A. J. Clin. Nutr. 68: 470S–475S, 1998.

2. *For general references for fatty acid and their affects on physical and mental function See:* Schmidt, M.A. **Smart Fats. How**

Dietary Fats and Oils Affect Mental, Physical and Emotional Intelligence. Frog Ltd, Berkeley, CA, 1997, p. 257; Udo, E. **Fats that Heal, Fats that Kill: The Complete Guide to Fats, Oils. Cholesterol and Human Health,** Alive Books, 1998; Udo, E. **Choosing the Right Fats: For Vibrant Health, Weight loss, Energy, Vitality.** Alive Books, 2002; Schmidt, M. M. **Brain-Building Nutrition: The Healing Power of Fats and Oils.** 2nd Ed., North Atlantic Books, June, 2001.

3. *Affects of types of Omega-3 fatty acids on brain function:* Connor, W.E. & Neuringer, M. The effects of n-3 fatty acid deficiency and repletion upon fatty acid composition and function of the brain and retina. In **Biological Membranes: Alterations in Membrane Structure and Function**. Alan R. Liss, Inc. 1988, pp. 275–294; Birch E.E., Hoffman, D.G. & Uauy. R. Dietary essential fatty acid supply and visual acuity development. Invest. Ophthalmol. Vis. Sci. 33(11):3242–3253, 1992; Stordy, J. Benefit of DHA supplement to dark adaptation in dyslexia. J. Clin. Nutr., 346:385, 1995; Stevens, L.J. et al. Essential fatty acid metabolism in boys as a possible cause of attention-deficit hyperactivity disorder. Am. J. Clin. Nutrition 65: 761–768, 1995; Stordy, J. Dark adaptation motor skills: docosahexaenoic acid and dyslexia. Am. J. Clin. Nutr. Supplement, 1997; Richardson, A. & Puri, B. A randomized double-blind, placebo controlled study of the effects of supplementation with highly unsaturated fatty acids on ADHD-related symptoms in children with specific learning difficulties. Progress in Neuro-Psychopharm & Biol. Psychiatry, 26(2): 233–239, 2002; Fox, D. The Speed of Life. Are your membranes gunky or runny? New Scientist 1 Nov. pp. 42–45, 2003; Hulbert, T. Life, death and membrane bilayers. J. Exp. Biol. 206:2303, 2003; Jensen, et al. Effects of maternal docosahexaenoic acid intake on visual function and neurodevelopment

in breastfed term infants. Am. J. Clin Nutrition, 82(1): 125–132, 2005.
4. *For an excellent fully referenced discussion of the Omega-3 oils see:* Larsen, H.R. Omega-3 Oils: Essential Nutrients. 28-03-2005. www.mercola.com/beef/omega3_oil .htm.
5. *Optimum ratios of DHA to EPA were derived from my original research using acupressure formatting for specific brain and body regions and muscle biofeedback.*
6. *For more information on the Omega-3 oils, factors that block conversion to EPA & DHA, and decreases in dietary levels see:* Schmidt, M.A. **Smart Fats. How Dietary Fats and Oils Affect Mental, Physical and Emotional Intelligence.** Frog Ltd, Berkeley, CA, 1997, p. 257; Schmidt, M. M. **Brain-Building Nutrition: The Healing Power of Fats and Oils.** 2nd Ed., North Atlantic Books, June, 2001; Larsen, H.R. Omega-3 Oils: Essential Nutrients. 28-03-2005. www.mercola.com/beef/omega3 _ oil. htm.
7. *For information on energy consumption and the executive functions see*: Munro, N.J. A model of the relationship among energy supply, energy demand and behaviour. In: Essman, W.B. (ed.) **Nutrients and Brain function.** Basil, Switzerland: Karger, , pp. 231–249, 1987; Schoenthaler, S.J. et al. Malnutrition and maladaptive behaviour: Two correlational analyses and a double blind placebo controlled challenge in five states. In Essman W.B. (ed.) **Nutrients and Brain Function.** Basil, Switzerland: Karger, p. 198–218, 1987; Zametkin, A.J. et al. Cerebral glucose metabolism in adults with Hyperactivity of Childhood Onset. New England. J. Med. 323(20):1361–1366, 1990; Magistretti, P.J. Brain energy metabolism. In: **Fundamental Neuroscience.** Eds. Zigmond, M.J. et al., Academic Press, N.Y., 1999, pp. 389–409; G.oldberg, E. **The Executive Brain. Frontal Lobes and the Civilized Mind.** Oxford University Press, New York, 2001.

Chapter 8. Nutrition for Optimum Mental Performance
1. *Information on amino acids precursors of neurotransmitters can be found in a variety of sources from popular books on nutrition to nutritional textbooks:* Davies, S. & Stewart, A. **Nutritional Medicine. The drug free guide to better family health.** Pan Books, London, 1987; Khalsa, D.S. *Brain Longevity.* Century Books, London, 1997; Werbach, M.R. **Nutritional Influences on Mental Illness: A Source Book of Clinical Research.** Thorsens Publishers Ltd, London, 2nd ed., 1991; Reavley, N. **Vitamins etc.** Bookman Press, Melbourne, 1999; *and these excellent recent books:* Carper, J. **Your Miracle Brain.** HarperCollins Publishers, N.Y. 2000; Sahelian, R. **Bottom Line's Mind Boosting Secrets. Natural Supplements that Enhance your Mind, Memory and Mood.** Bottom Line Books, St. Martin's Press, 2005; Holford, P. & Colson, D. **Optimum Nutrition for your Child's Mind: Maximising your Child's Potential.** Piatkus, 2006, p. 294.
2. *The role of amino acid supplementation for enhanced mental function has received some attention:* Liebman, H.R. Tyrosine and Stress: **Human and Animal Studies. Food Components to Enhance Performance.** National Academy Press, pp.277–299, 1994; Tyrosine: Food supplement or therapeutic agent? J. Nutr. & Environ. Med. 8:349–359, 1998; Shiah, I.S. & Yatham. GABA functions in mood disorders: An update and critical review. Nature Life Science, 63(15):1289–1303, 1998; Tyrosine improves working memory in a multitasking environment. Pharmcol. Biochem. & Behav. 64(3):495–500, 1999; Deijen et al. Tyrosine improves cognitive performance and reduces blood pressure in cadets. Brain Res. Bulletin, 48(2):203–209, 1999; Young, S.N. & Leyton, M. The role of serotonin in human and social interactions: Insight from altered tryptophan levels. Pharmacol. Biochem. & Behav. 71:857–865, 2002; Rodgers, R.D. et al. Tryptophan depletion

alters the decision-making of healthy volunteers through altered processing of reward cues. Neuropsycho-pharmacol. 28:153–162, 2003; Turner, E. et al. Serotoninalacarte: Supplementation with the serotonin precursor 5-Hydroxytryptophan. July 13 (Epub), 2005.

3. *The role of dopamine in reward and punishment and how dopamine deficiency may lead to cognitive problems and addiction is covered in:* Miller, D. & Blum, K. **Overload. Attention Deficit Disorder and the Addictive Brain.** Andrews & McMeel, Kansas City, MO, pp. 55–67, 1996; Valkow, N. et al. Decreased striatal dopaminergic responsiveness in detoxified cocaine-dependent subjects. Nature 386:830–833, 1997; Blum, K., Holder, J. The Reward Deficiency Syndrome. The American College of Addictionology & Compulsive Disorders, Amereol, Ltd, 2002.

4. *The ratios of Omega-6 to Omega-3 fatty acids and types of fish oils are reported in:* Connor, W.E. & Neuringer, M. The effects of n-3 fatty acid deficiency and repletion upon fatty acid composition and function of the brain and retina. In **Biological Membranes: Alterations in Membrane Structure and Function**. Alan R. Liss, Inc. 1988, pp. 275–294; Birch E.E., Hoffman, D.G. & Uauy, R. Dietary essential fatty acid supply and visual acuity development. Invest. Ophthalmol. Vis. Sci. 33(11): 3242–3253, 1992; Larsen, H.R. Omega-3 Oils: Essential Nutrients. 28-03-2005. **www.mercola.com/beef/omega3_oil. htm**. *In the following paper Japanese researchers showed that the imbalance between Omega-6 & Omega-3 fatty acids may affect learning and retinal functions (and hence vision):* Yoshida, S., Sato, A. & Okuyama, H. Pathophysiological effects of dietary essential fatty acid balance on neural systems. Jpn J. Pharmcol. 77:11–22, 1998.

5. *The role fatty acids may play in mental function is highlighted by studies showing their deficiency may lead to overt cognitive*

dysfunction: Stordy, J. Benefit of DHA supplement to dark adaptation in dyslexia. Lancet, 346:385, 1995; Stevens, L.J. et al. Essential fatty acid metabolism in boys with attention-deficit hyperactivity disorder. Am. J. Clin. Nutr., 65:761–768, 1995; Stordy, J. Dyslexia, attention deficit hyperactivity disorder, dyspraxia – do fatty acids help? Dyslexia Rev. 9(2), 1997; Richardson, A.J. et al. Is developmental dyslexia a fatty acid deficiency syndrome? Proc. Nutr. Society, Ann. Conf., 1998; Richardson, A.J. et al. Fatty acid deficiency signs predict the severity of reading and related difficulties in dyslexic children. Prostaglandins Leukotr. & Essential Fatty Acids, 63: 69–74, 2000; Stordy, B.J. Dark adaptation, motor skills, docosahexaenoic acid and dyslexia. Am. J. Clinical Nutr. 71(1):323S-326S, 2000; Richardson, A.J. & Montgomery, P. The Oxford-Durham study: A randomized controlled trial of dietary supplementation with fatty acids in children with developmental coordination disorder. Pediatrics, 115:1360–1366, 2005; Jensen, et al. Effects of maternal docosahexaenoic acid intake on visual function and neurodevelopment in breastfed term infants. Am. J. Clin Nutrition, 82(1):125–132, 2005; Richardson, A.J. et al. Reduced behavioural and learning problems in children with specific learning difficulties after supplementation with highly unsaturated fatty acids. European J. Neuroscience 12: (Suppl.11): 296, 2006; *The role that deficiency of the Omega-3 fatty acids EPA & DHA play in ADHD are clearly demonstrated by recent studies:* Richardson, A.J. A randomized double-blind, placebo-controlled study of the effects of supplementation with highly unsaturated fatty acids on ADHD-related symptoms in children with specific learning problems. Prog. Neuropharm. & Biol. Psychiatry, 26(2):233–239, 2002; Young, G.S., Conquer, J.A. & Thomas, R. Effect of randomized supplementation with high dose olive, flax or fish oils on serum phospholipid fatty acid levels in adults with attention deficit hyperactivity disorder. Reprod. Nutr. Dev. 45(5):549–558, 2005; Richardson, A.J. Omega-3 fatty acids in ADHD and related neurodevelopmental disorders. Int. Rev Psychiatry 18(2):155–172, 2006 (*an excellent review of this topic*).

6. *The effect of fatty acids on learning an behavioural problems is also summarised in Dr. Jacqueline Stordy's recent book:* Stordy, B.J. and Nicoll, M.J., **The LCP Solution: The Remarkable Nutritional Treatment for ADHD, Dyslexia and Dyspraxia**. Ballatine Publishing Group, London, 2000, p. 192.

7. *Role of B-Vitamins in brain function:* Mindell, E. **Earl Mindell's Vitamin Bible**. Warner Books September 2004; Bernstein, A.L. Vitamin B_6 in Clinical Neurology. Ann. N.Y. Acad. Sci. 585: 250–260, 1990; Guilarte, T.R. Vitamin B_6 and cognitive development: Recent research findings from humans and animals. Nutr. Reviews 51(7):193–198, 1993; Reavley, N. **Vitamins etc**. Bookman Press, Melbourne, 1998, pp. 59–127; *B-Vitamins, cognition and aging:* A review. J. Gerontology: Psych. Sci. 56B(6):P327–P339, 2001; Ullegaddi, R., Powers, H.J. & Gariballa, S.E. B-group vitamin supplementation mitigates oxidative damage after acute ischaemic stroke. Clinical Sci. 107:477–484, 2004.

8. *The role of Nootropic herbs that enhance brain function*: Khalsa, D.S. **Brain Longevity**. Century Books, London, 1997, pp. 259–262; Sharma, R. Efficacy of Bacopa monniera in revitalizing intellectual functions in children. J. Res. Edu. Ind. Med. Jan–June:1-10, 1987; Rigney, U., Kimber, S. & Hindmarch, I. The effects of acute doses of standardized Gingko biloba extract on memory and psychomotor performance in volunteers. Phytother. Res. 13:408–415, 1999; Vohoara, D. Pal, S.N. & Pillai, K.K. Protection from phenytoin cognitive deficit by Bacopa monniera, a reputed Indain nootropic plant. J. Ethnopharmacology 71: 383–390, 2000; Stough, C. et al. The

chronic effects of an extract of Bacopa monniera (Brahmi) on cognitive function in healthy human subjects. Psychopharmocolgy 17 April: 1–8, 2001; Das, A. et al. A comparative study in rodents of standardized extracts of Bacopa monniera and Gingko biloba on anticholinesterase and cognitive enhancing activities. Pharmaocol, Biochem. & Behav. 73:893–900, 2002; Kar Chowdhuri, D. et al. Antistress effects of bacosides of Bacopa monneri: modulation of Hsp 70 gene expression, superoxide dismutase and cytochrome P-450 activity in the rat brain. Phytother. Res. 16:639–645, 2002; Roodenrys, S. et al. Chronic effects of Brahmi (bacopa monniera) on human memory. Neuropsychopharmocology 27 (2):279–281, 2002; Rai, D. et al. Adaptogenic effect of Bacopa monniera (Brahmi). Pharmocol. Biochem. & behave. 75:823–830, 2003; Russo, A. et al. Free radical scavenging capacity and protective effect of Bacopa monniera L. on DNA damage. Phytother. Res. 17:870–875; 2003.

9. *Table 7. A list of nutrients and nootropic herbs to optimise mental function was derived from the primary research literature, and a number of consolidated sources such as the books given below:* Dean, W. & Morgenthaler, J. **Smart Drugs & Nutrients.** B & J Publications, Santa Cruz, CA, 1990; Werbach, M.R. **Nutritional Influences on Mental Illness: A Source Book of Clinical Research.** Third Line Press, New 2nd Ed., Tarzana, CA, 1993; Khalsa, D.S. ***Brain Longevity.*** Century Books, London, 1997; Schmidt, M.A. **Smart Fats. How Dietary Fats and Oils Affect Mental, Physical and Emotional Intelligence.** Frog Ltd, Berkeley, CA, 1997, p. 257; Schmidt, M. M. **Brain-Building Nutrition: The Healing Power of Fats and Oils.** 2nd Ed., North Atlantic Books, June, 2001; Mindell, E. **Earl Mindell's Vitamin Bible.** Warner Books September 2004; Holford, P. **Optimum Nutrition for the Mind.** Basic Health Publications, 2004; Sahelian, R. **Mind Boosting Secrets. Natural Supplements that Enhance your Mind, Memory and**

Mood. Bottom Line Books, St. Martin's Press, 2005; Holford, P. & Colson, D. **Optimum Nutrition for your Child's Mind: Maximising your Child's Potential.** Piatkus, 2006, pp. 294.
10. *For examples of studies that have shown that therapeutic doses of nutrients far exceed RDA's see:* Reading, C.M. Family Tree connection: How your past can shape your future health. A lesson in orthomolecular medicines. J. Orthomol. Med. 3(3):123–134, 1988, *For the case studies with B_3 & B_6; and Appendix C, In*: Werbach, M.R. **Nutritional Influences on Mental Illness: A Source Book of Clinical Research.** 2nd Ed., Third Line Press, Tarzana, CA, 1991, p.4 66. *This table compares RDA's and the pharmacologic dosage range for various nutrients.*

Chapter 9. Nutriceuticals for Optimising Mental Performance
1. *There are now a number of books discussing the value of vitamin, mineral and herbal supplementation for mental performance:* Dean, W. & Morgenthaler, J. *Smart Drugs & Nutrients.* B & J Publications, Santa Cruz, CA, 1990; Schmidt, M.A. **Smart Fats. How Dietary Fats and Oils Affect Mental, Physical and Emotional Intelligence.** Frog Ltd, Berkeley, CA, 1997, p. 257; Schmidt, M. M. **Brain-Building Nutrition: The Healing Power of Fats and Oils.** 2nd Ed., North Atlantic Books, June, 2001; Stordy, B.J. and Nicoll, M.J., **The LCP Solution: The Remarkable Nutritional Treatment for ADHD, Dyslexia and Dyspraxia.** Ballatine Publishing Group, London, 2000, p.192; Carper, J. **Your Miracle Brain.** HarperCollins Publishers, N.Y. 2000; Holford, P. **Optimum Nutrition for the Mind.** Basic Health Publications, 2004; Sahelian, R. **Mind Boosting Secrets. Natural Supplements that Enhance your Mind, Memory and Mood.** Bottom Line Books by arrangement with St. Martin's Press, 2005.
2. *Nutriceuticals for maintaining and enhancing brain function are relatively new with most being introduced within the past 5 years.*

3. *The following websites present information on the different nutriceuticals for enhancing brain function mentioned in the text*: www.vavax.com; www.focusfactor.com; www.learningfactor.com; www.northstarnutritionals.com; www.nutriwest.com and www.thinkingadvantage.com.

Appendix 1 & 2: Vitamin & Mineral Sources, Doses, Functions and Deficiency/Excess Disorders.

1. *These tables were derived from the primary literature and the complied sources below*: Bloomfield, M.M. & Lawrence, J.S. **Chemistry and the Living Organism**. 6th Ed., John Wiley & Sons, N.Y., 1996, p. 475; **Dietary Reference Intakes: Applications in Dietary Assessment, Institute of Medicine.** National Academy Press, Washington, DC, 2001; **Dietary Reference Intakes for Vitamin C, Vitamin E, Selenium and Carotenoids, Institute of Medicine, National Academy Press, Washington, D.C, 2001; Dietary Reference Intakes for Vitamin A, Vitamin K, Arsenic, Boron, Chromium, Copper, Iodine, Iron, Manganese, Molybenum, Nickel, Silicon, Vanadium and Zinc, Institute of Medicine.** National Academy Press, Washington, DC, 2001.

2. *The Therapeutic Dosage Range, the amount of various nutrients required to optimise physiological functions, as opposed to the RDAs that only prevent overt nutritional deficiency disease were taken from a number of books, the primary literature and the summary in*: Werbach, M.R. **Nutritional Influences on Illness. A Source Book of Clinical Research.** Thorson's Publishing Group, London, 1987, p. 466.

Appendix 3: Nutriceuticals for Optimizing Mental Performance

1. *Information contained here is derived from the research and development of ThinkingAdvantage and over 15 years of clinical practice using Applied Neurology direct muscle biofeedback to*

evaluate nutritional products to enhance learning and brain function.

2. *Further information can be found on the ThinkingAdvantage website:* **www.thinkingadvantage.com.** For readers of this book, the nutriceutical ThinkingAdvantage for Brain Integration may be ordered at a 10% discount by going to **www.thinkingadvantage.com** click on New Customer Registration and enter Promotion Code 777.

3. *Further information on Brain Integration can be found on the website*: **www.Leapbrainintegration.com,** *and in my book*: Krebs, C.T. & Brown, J. **A Revolutionary Way of Thinking. From near Fatal Accident to a New Science of Healing.** Hill of Content Publishing, 1998, chapters 5 to 9; *Available from:* www.kinesiologyshop.com.au & www.amazon.com.

APPENDICES 1 & 2

Vitamins and Minerals: their Sources, Recommended Dietary Allowance (RDA), Therapeutic Dosage Range (TDR), Functions & Deficiency Symptoms and Disorders

Appendix 1: Table of Vitamins, Sources, Doses, Functions and Deficiency Symptoms and Disorders

Vitamins	Storage & Comment	Major Food Sources	RDA or DV[a] versus Therapeutic Dosage Range[b] (TDR)	Functions	Deficiency Symptoms and Disorders
Fat-soluble:	All require bile salts and some dietary lipids for adequate absorption				
Vitamin A	Formed from provitamin beta-carotene (and other provitamins) in GI tract. Stored in liver. Yellow and green vegetables; sources of vitamin A include liver and milk.	Sources of carotene and other provitamins include: carrots, green leafy vegetables, sweet potatoes, pumpkin, winter squash, apricots, cantaloupe, fortified margarine. Retinol: liver, butter, whole milk, cheese, egg yolks.	RDA of 1mg or ~5000 IU (800 µg of retinol[c]) Versus TDR of 2 to 7 mg or 10,000 to 35,000 IU	Maintains general health and vigour of epithelial cells. Beta-carotene acts as an antioxidant to inactivate free radicals. Formation of photopigments, light-sensitive chemicals in photoreceptors of retina. Involved in transport of nutrients across cell membranes. Aids in growth of bones and teeth, apparently by helping to regulate activity of osteoblasts and osteoclasts.	Deficiency results in atrophy and keratinization of epithelium, leading to dry skin and hair; increased incidence of ear, sinus, respiratory, urinary, and digestive system infections; drying of cornea; and skin sores. **Night blindness** or decreased ability for dark adaptation. Slow and faulty development of bones and teeth.

Appendix I: Continued

Vitamins	Storage & Comment	Major Food Sources	RDA or DV[a] versus Therapeutic Dosage Range[b] (TDR)	Functions	Deficiency Symptoms and Disorders
Fat-soluble:	All require bile salts and some dietary lipids for adequate absorption				
Vitamin D (Note due to lack of response to sunlight, many people in industrialised countries today are Vitamin D deficient.)	Sunlight converts 7-dehydrocholesterol in the skin to vitamin D3 (cholecalciferol). A liver enzyme then converts cholecalciferol to 25-hydroxycholecalciferol. A second enzyme in the kidneys converts 25-hydroxy cholecalciferol to calcitriol (1,25-dihydroxycalciferol), which is the active form of vitamin D. Most excreted via bile.	Provitamins in skin activated by sunlight. Fish oils, egg yolk, liver, margarine and fortified dairy products.	RDA of 10 μg or 400 IU of cholecalciferol. Versus TDR of 25 to 100 mg or 1000 to 4000 IU	Essential for absorption and utilization of calcium and phosphorus from GI tract. Works with parathyroid hormone (PTH) to maintain Ca^{+2} homeostasis. Increases absorption of calcium and promotes ossification of bones and teeth.	Defective utilization of calcium by bones leads to **rickets** in children and osteomalacia in adults, and is involved in osteoporosis. Possible loss of muscle tone.

Appendix 1: Continued

Vitamins	Storage & Comment	Major Food Sources	RDA or DV[a] versus Therapeutic Dosage Range[b] (TDR)	Functions	Deficiency Symptoms and Disorders
Fat-soluble:	All require bile salts and some dietary lipids for adequate absorption				
Vitamin E (tocopherols) (Recent studies showing negative effects of Vit E supplements were all based upon studies using synthetic Vit E, not naturally occurring tocopherols.)	Stored in liver, adipose tissue, and muscles. Only occurs naturally in foods as RRR-alpha tocopherol, not the stereoisomers used in fortified foods and supplements.	Vegetable oil, liver, margarine (with no trans fatty acids), green and leafy vegetables, wheat germ, egg yolk, butter and whole-grain products.	RDA of 10 to 15 mg or 150 to 225 IU (8mg) of alpha-tocopherol Versus TDR of 20 to 100 mg or 300 to 1500 IU	Antioxidant protecting Vitamins A & C, and monounsaturated fats that help form and maintain cell structures, especially membranes. Involved in formation of DNA, RNA, and red blood cells. Promotes wound healing, the normal structure and functioning of the nervous system, and prevent scarring. Acts as an antioxidant to inactivate free radicals, & helps protect liver from toxic chemicals such as carbon tetrachloride.	Deficiency results in abnormal structure and function of mitochondria, lysosomes, and plasma membranes. Deficiency also causes increased fragility of red blood cells. A possible consequence is hemolytic anaemia. Deficiency also causes muscular dystrophy in monkeys and sterility in rats.

Appendix 1: Continued

Vitamins	Storage & Comment	Major Food Sources	RDA or DV[a] versus Therapeutic Dosage Range[b] (TDR)	Functions	Deficiency Symptoms and Disorders
Fat-soluble:	All require bile salts and some dietary lipids for adequate absorption				
Vitamin K	Produced by intestinal bacteria. Stored in liver and spleen.	Dietary sources include spinach, cauliflower, cabbage, liver.	RDA of 70 µg (60 µg) Versus TDR of 70 to 150 µg	Coenzyme essential for synthesis of several clotting factors by liver, including prothrombin.	Delayed clotting time results in excessive bleeding. In excess: haemolytic anaemia and liver damage.
Water-soluble:	Absorbed along with water in GI tract and dissolved in bloody fluids				
Vitamin B$_1$ (thiamine)	Rapidly destroyed by heat. Not stored in body. Excess intake eliminated in urine.	Sources include whole-grain products, eggs, pork, nuts, liver, yeast, liver, meat, fortified grain products, and legumes.	RDA of 1.1 to 1.5 mg Versus TDR of 10 to 200 mg	Acts as coenzyme for many different enzymes that break carbon-to-carbon bonds and are involved in carbohydrate metabolism of pyruvic acid to CO_2 and H_2O. Essential for synthesis of acetylcholine, and thought to have independent role in nerve conduction.	Improper carbohydrate metabolism leads to build-up of pyruvic and lactic acids and insufficient production of ATP for muscle and nerve cells. Deficiency leads to: (1) **beri beri** – partial paralysis of smooth muscle of GI tract, causing digestive disturbances; skeletal muscle paralysis; and

Appendix 1: Continued

Vitamins	Storage & Comment	Major Food Sources	RDA or DV[a] versus Therapeutic Dosage Range[b] (TDR)	Functions	Deficiency Symptoms and Disorders
Water-soluble:	Absorbed along with water in GI tract and dissolved in bloody fluids				
Vitamin B₁ continued					atrophy of limbs; heart failure. I(2) **polyneuritis** – due to degeneration of myelin sheaths; impaired reflexes, impaired sense of touch, stunted growth in children, and poor appetite.
Vitamin B₂ (riboflavin)	Not stored in large amounts in tissues. Most is excreted in urine. Small amounts supplied by bacteria of GI tract.	Dietary source include yeast, liver, beef, lamb, eggs, yeast, wholegrain, products, asparagus, peas, beets, peanuts milk, yogurt, cottage cheese, and fortified grain products.	RDA of 1.3 to 1.7 mg Versus TDR of 10 to 50 mg	Coenzyme in oxidation reactions and component of certain coenzymes (for example, FAD and FMN) in carbohydrate and protein metabolism, especially in cells of eye, integument, mucosa of intestine, blood.	Deficiency may lead to improper utilization of oxygen resulting in blurred vision, cataracts, and corneal ulcerations – often felt as sand or grit in eyes. Also dermatitis and cracking of skin, lesions of intestinal mucosa, and development on one type of anaemia.

Appendix 1: Continued

Vitamins	Storage & Comment	Major Food Sources	RDA or DV[a] versus Therapeutic Dosage Range[b] (TDR)	Functions	Deficiency Symptoms and Disorders
Water-soluble:	Absorbed along with water in GI tract and dissolved in bloody fluids				
Vitamin B$_3$ (niacin or nicotinamide)	Derived from amino acid tryptophan.	Sources include yeast, meats, liver, fish, poultry, whole-grain products, peas, beans, nuts, peanuts and fortified grain products.	RDA of 15 to 18 mg Versus TDR of 100 to 6000 mg	Essential component of NAD and NADP, coenzymes in oxidation-reduction reactions. In lipid metabolism, inhibits productions of cholesterol and assists in triglyceride breakdown.	atrophy of limbs; heart Principal deficiency is **pellagra**, characterized by dermatitis, diarrhoea, and psychological disturbances including dementia.
Vitamin B$_6$ (pyridoxine)	Synthesized by bacteria of GI tract. Stored in liver, muscle, brain.	Dietary sources include meat, poultry, liver, fish, shellfish, yeast, yogurt, tomatoes, yellow corn, spinach, green leafy vegetables, whole-grain products and legumes.	RDA of 1.6 to 2.0 mg Versus TDR of 10 to 200 mg	Essential coenzyme for normal amino acid and fatty acid metabolism. Assists production of antibodies and needed for synthesis of heme. Major coenzyme in production of brain neurotransmitters like serotonin, noradrenalin and dopamine.	Most common deficiency symptoms: Convulsions in infants, and dermatitis of eyes, nose, mouth and other skin disorders in adults. Other symptoms are retarded growth and nausea. Loss of brain integration due to lower levels of essential neurotransmitters.

Appendix 1: Continued

Vitamins	Storage & Comment	Major Food Sources	RDA or DV[a] versus Therapeutic Dosage Range[b] (TDR)	Functions	Deficiency Symptoms and Disorders
Water-soluble:	Absorbed along with water in GI tract and dissolved in bloody fluids				
Vitamin B_{12} (cyanoco-balamin)	Only B vitamin not found in vegetables; only vitamin containing cobalt. Absorption from GI tract depends on intrinsic factor secreted by gastric mucosa.	Sources include meat, liver, kidney, milk, eggs, cheese, poultry, fish, eggs, shellfish, and dairy products.	RDA of 2.0 to 2.8 μg Versus TDR of 10 to 200 μg 10 to 30% of older people malabsorb food-bound vitamin B_{12} so those over 50 should eat fortified foods or take a vitamin supplement containing B_{12}.	Coenzyme necessary for red blood cell formation, formation of the amino acid methionine, entrance of some amino acids into Krebs cycle, and manufacture of choline (used to synthesize acetylcholine, an important neurotransmitter in brain and muscle function).	**Pernicious anaemia,** neuropsychiatric abnormalities (ataxia, memory loss, weakness, personality and mood changes, and abnormal sensations), and impaired osteoblast activity.
Pantathenic acid	Stored primarily in liver and kidneys. Some produced by bacteria of GI tract.	Dietary sources include kidney, liver, yeast, meats, milk, egg yolk, green leafy vegetables, whole-grain products and legumes.	RDA of 4 to 7 mg Versus TDR of 50 to 1000 mg	Part of coenzyme-A needed to transfer acetyl group from pyruvic acid into Krebs Cycle respiration; conversion of lipids and amino acids into glucose; and synthesis of cholesterol and steroid hormones.	Neuromotor, digestive and cardiovascular disorders. Experimental deficiency tests indicate fatigue, muscle spasm, neuro-muscular degeneration, and insufficient production of adrenal steroid hormones.

Appendix I: Continued

Vitamins	Storage & Comment	Major Food Sources	RDA or DV[a] versus Therapeutic Dosage Range[b] (TDR)	Functions	Deficiency Symptoms and Disorders
Water-soluble:	Absorbed along with water in GI tract and dissolved in bloody fluids				
Folic acid (folate, folacin)	Synthesized by bacteria of GI tract.	Dietary sources include green leafy vegetables, broccoli, asparagus, breads, dried beans, legumes, liver, and citrus fruits.	RDA of 400 µg Versus TDR of 400 to 2000 µg	Coenzyme in nucleic acid synthesis producing purines and pyrimidines built into DNA and RNA. Essential for normal production of red and white blood cells and amino acid metabolism.	Production of abnormally large red blood cells (**macrocytic anaemia**). Inhibition of cell division with higher risk of neural tube defects in babies born to folate-deficient mothers. Digestive disorders.
Biotin	Synthesized by bacteria of GI tract.	Dietary sources include yeast, kidneys, liver, egg yolk, milk, most fresh vegetables.	RDA of 30 to 200 µg Versus TDR of 300 to 3000 µg	Coenzyme in fatty acids and purine synthesis, and amino acid metabolism and glycogen formation.	Mental depression, anorexia, muscular pain, dermatitis, fatigue, nausea.

Appendix 1: Continued

Vitamins	Storage & Comment	Major Food Sources	RDA or DV[a] versus Therapeutic Dosage Range[b] (TDR)	Functions	Deficiency Symptoms and Disorders
Water-soluble:	Absorbed along with water in GI tract and dissolved in bloody fluids				
Vitamin C (Ascorbic acid)	Rapidly destroyed by heat. Some stored in glandular tissue and plasma.	Sources include citrus fruits, tomatoes, leafy vegetables, potatoes, and cabbage.	RDA of 60 to 90mg Versus TDR of 100 to 10,000 mg	Although exact role not fully understood, promotes many metabolic reactions, especially collagen and connective tissue synthesis. As coenzyme, combines with poisons, rendering them harmless, and is an important antioxidant, especially in epithelial tissues, and promotes wound healing. Enhances antibody action, and role in cancer prevention under investigation.	**Scurvy,** bleeding gums, loosened teeth, swollen joints and many symptoms related to poor connective tissue growth and repair. Poor wound healing, bleeding (vessel walls fragile because of connective tissue degeneration), anaemia and retardation of growth. Deficiency appears to reduce immune system response to viruses and bacteria, and may play a role in cancer.

[a] RDAs given as ranges, because women and men often require different amounts and American and European standards differ for the same nutrient, with higher concentrations in the United States.
[b] Therapeutic Dosage Range derived from the literature and *Nutritional Influences on Mental Illness: A Source Book of Clinical Research*, by Mervyn Werbach.
[c] Calculation of vitamin A is complex and usually now expressed as μg of retinol activity equivalents (RAEs), as the dietary source of vitamin A are a group of carotenoids (carotenes) that are then converted to retinol in the liver as the body needs this nutrient.

Appendix 2: Table of Minerals, Sources, Doses, Functions and Deficiency Symptoms and Disorders

Vitamins	Storage & Comment	Major Food Sources	RDA or DV[a] versus Therapeutic Dosage Range[b] (TDR)	Functions	Deficiency/Excess Symptoms and Disorders
Macro-Minerals Found in the Human Body					
Calcium	Most abundant cation (positively charged ion) in the body. About 99% is stored in bone and teeth in combination with phosphorus. Controlled by calcitonin and parathyroid hormone, but absorption occurs only in presence of vitamin D. Excess is excreted in faeces and urine.	Dietary sources include egg yolk, shellfish, green leafy vegetables, milk and milk products and many nuts with almonds having several times more bio-available calcium than cow's milk.	RDA of 800 to 1200 mg Versus TDR of 1000 to 1500 mg	Formation of bones and teeth, blood clotting, normal muscle and nerve activity, cellular mobility, glycogen metabolism, and absorption of B group vitamins. Also required for synthesis and release of neurotransmitters, and a major ion controlling intracellular activities, including programmed cell death or apotosis.	As a macro-mineral that is abundant in many common food sources, calcium does not commonly create a deficiency disease. However, deficiency in the diet is associated with **osteomalacia** in adults and **osteoporosis**, especially in the elderly.
Phosphorus	About 80% is found in bones and teeth with remainder distributed in muscles, brain, and blood.	Dietary sources include meat, fish, dairy products, nuts, cereals, legumes and poultry.	RDA of 700 to 1250 mg Versus TDR of ? Unknown	Formation of bones and teeth. Phosphates constitute a major part of the buffer system of blood that regulates the pH.	Because it is a common macro-mineral it is seldom deficient and because it is easily excreted, excess is generally not a problem.

Appendix 2: Continued

Vitamins	Storage & Comment	Major Food Sources	RDA or DV[a] versus Therapeutic Dosage Range[b] (TDR)	Functions	Deficiency/Excess Symptoms and Disorders
Macro-Minerals Found in the Human Body					
Phosphorus continued	Participates in more functions than any other mineral. Controlled by calcitonin and parathyroid hormone and excess is excreted in faeces and urine.			Plays an important role in muscle contraction and nerve activity, and is a co-factor for many enzymes. As an important part of ATP, is involved in energy transfer in the cell, and component of DNA and RNA.	
Potassium	Principal cation (K⁺) in intracellular fluid. Excess excreted in urine. Normal food intake supplies required amounts.	Dietary sources, oranges, dried fruits, bananas, meats, potatoes, peanuts butter and coffee.	RDA of 1800 to 5600 mg Versus TDR of ? Unknown	Functions in nerve and muscle action by creating nerve impulse and muscle action potential when neurotransmitters released at nerve synapse open sodium and potassium gates.	Because it is such a common macro-mineral, potassium is seldom deficient even in junk food diets. Diuretics may cause deficiency and supplementation is often required.

APPENDICES 1 & 2

Appendix 2: Continued

Vitamins	Storage & Comment	Major Food Sources	RDA or DV[a] versus Therapeutic Dosage Range[b] (TDR)	Functions	Deficiency/Excess Symptoms and Disorders
Macro-Minerals Found in the Human Body					
Sulfur	Constituent of many proteins (such as insulin), electron carriers in Krebs Cycle oxidative phosphorylation, and some vitamins (thiamine and biotin). Excreted in urine.	Dietary sources include beef, liver, lamb, fish, poultry, eggs, cheese and beans.	RDA of ? mg Versus TDR of 500 to 1000 mg	Important structural element of many proteins, as sulphur cross-bridges give many molecules their shape (e.g. is responsible for wavy and kinky hair). As components of hormones and vitamins, regulates various body activities. Needed for ATP production by aerobic cellular respiration.	Generally plentiful in most diets, but for some people with genetic defects in certain sulfur containing enzymes and proteins more sulphur may be required than is found in their diets.
Sodium	Most abundant cation (Na^+) in extracellular fluids; some found in bones. Excreted in urine and perspiration.	Normal intake of NaCl (table salt) supplies more than the required amounts, also in cured meats, meats, milk and olives.	RDA of 1100 to 3300 mg Versus TDR of ? Unknown	Strongly affects distribution of water through osmosis. Part of bicarbonate buffer system controlling pH in the body. Functions in conjunction with potassium in creating nerve impulses and muscle action potentials.	Because it is such a common macro-mineral, sodium is seldom deficient even in junk food diets. In fact, sodium excess from junk foods and excessive use of table salt is more the problem affecting blood pressure and osmotic balance.

Appendix 2: Continued

Vitamins	Storage & Comment	Major Food Sources	RDA or DV[a] versus Therapeutic Dosage Range[b] (TDR)	Functions	Deficiency/Excess Symptoms and Disorders
Macro-Minerals Found in the Human Body					
Chloride	Principal anion (Cl⁻) in extracellular fluid. Excess excreted in urine.	Normal intake of NaCl supplies required amounts, but also in meats, milk and eggs.	RDA of 500 to 2500 mg Versus TDR of ? Unknown	Plays role in acid-base balance of blood, water balance, and formation of hydro-chloric acid (HCl) in stomach.	Because it is such a common macro-mineral, chloride is seldom deficient even in junk food diets.
Magnesium	Important cation (Mg^{+2}) in intracellular fluid. Excess excreted in urine and faeces.	Widespread in various food, such as green leafy vegetables, nuts, seafood, milk, and whole-grain cereals.	RDA of 300 to 350 mg Versus TDR of 300 to 800 mg	Constituent of many coenzymes and is itself an important co-factor in many enzymes, including those making brain neurotransmitters. Participates in bone formation. Required for normal nerve and muscle function.	Deficiency often leads to muscle cramps, as magnesium regulates muscle contraction. Deficiency can result in children showing excessive fidgeting, psychomotor instability and learning difficulties. It is also an important co-factor in neurotransmitters linked to aggression, and deficiency is associated with Type A behaviour and aggression.

Appendix 2: Continued

Vitamins	Storage & Comment	Major Food Sources	RDA or DV[a] versus Therapeutic Dosage Range[b] (TDR)	Functions	Deficiency/Excess Symptoms and Disorders
Micro-Minerals Found in the Human Body					
Iron	Widespread deficiency due to presence of phytates in food binding it and loss in faeces. About 66% of iron is found in hemoglobin in blood. Losses by shedding of hair, epithelial cells, and mucosal cells, and in sweat, urine, faeces, bile and blood lost during menstruation.	Dietary sources are meat, shellfish, heart, liver, clams, oysters, spinach, dates, egg yolk, beans, legumes, dried fruits, nuts and cereals. Most supplementary forms are difficult to absorb.	RDA of 10 to 18 mg Versus TDR of 10 to 50 mg	Main structural element in hemoglobin molecule that reversibly binds O_2 in blood. Component of cytochromes involved in electron transport chain of cellular respiration providing ATP for nerve and muscle action as well as all cell activities. Major co-factor for enzymes making all of the primary neurotransmitters in the brain.	**Anaemia** is the most commonly recognised deficiency problem. Deficiency has major affects on learning and behaviour due to its role as a co-factor for the enzymes producing the major neurotransmitters controlling brain function. Deficiency leads to reduced IQ, cognitive functions and lethargy.
Zinc	Along with iron, one of most widespread nutrient deficiencies due to loss from soils, and presence	Widespread in many foods, whole-grain cereals, eggs, legumes and especially red meats and seafoods.	RDA of 12 to 15 mg Versus TDR of 20 to 100 mg	Co-factor for over 300 enzymes, e.g. carbonic anhydrase important in carbon dioxide metabolism and peptidases, it is	Deficiency, even marginal deficiency has major affects on learning and behaviour especially short-term memory and executive

Appendix 2: Continued

Vitamins	Storage & Comment	Major Food Sources	RDA or DV[a] versus Therapeutic Dosage Range[b] (TDR)	Functions	Deficiency/Excess Symptoms and Disorders
Micro-Minerals Found in the Human Body					
Zinc continued	of phytates in food that bind zinc and iron in the gut and remove them from the body.			involved in protein digestion. Necessary for normal growth and wound healing, normal taste sensations and appetite, and normal sperm counts in males. Co-factor involved in producing major brain neurotransmitters.	functions involved in thinking and reasoning. If deficiency is extreme, psychotic disorders may result, e.g. schizophrenia. Deficiency also results in suppressed immune function, and reduced liver detoxification, especially of heavy metals.
Iodide	Due to iodine deficiency of many soils, iodide is a common deficiency. Excreted in urine.	Dietary sources are seafood, iodized salt, and vegetables grown in iodine-rich soils.	RDA of 150 µg Versus TDR of 150 to 1000 µg	Required by thyroid gland to synthesize thyroid hormones, which regulate metabolic rate. The thyroid hormones, thyroxins, are tri- and tetra-iodates, meaning they need 3 or 4 iodine atoms.	Deficiency causes the thyroid tissue to increase in a vain attempt to produce more thyroxin, resulting in a **goiter**, or swelling at the base of the neck, and these are common in areas with low iodine levels in the soil.

Appendix 2: Continued

Vitamins	Storage & Comment	Major Food Sources	RDA or DV[a] versus Therapeutic Dosage Range[b] (TDR)	Functions	Deficiency/Excess Symptoms and Disorders
Micro-Minerals Found in the Human Body					
Manganese	Some stored in liver and spleen. Most excreted in faeces. As a trace element, very little is needed	Dietary sources liver, and organ meats, and vegetables.	RDA of 2.5 to 5 mg Versus TDR of 5 to 50 mg	Activates several enzymes needed for hemoglobin synthesis, urea formation, growth, reproduction, lactation, bone formation, and possibly production and release of insulin, and inhibition of cell damage.	As a trace element, little is needed so it is seldom deficient in most diets. Excess has been associated with increased aggressiveness
Copper	Some stored in liver and spleen. Most excreted in the faeces. When zinc is deficient, then copper is often in excess.	Dietary sources include meats, liver, fish, shellfish, whole-grain flour, eggs, beans, beets, spinach, legumes, nuts, grapes, and asparagus.	RDA of 2 to 3 mg Versus TDR of 2 to 4 mg	Required for coenzymes in electron transport chain making ATP; enzymes necessary for melanin formation; and needed with iron for synthesis of hemoglobin.	Deficiency reduces energy production and may lead to anaemia. Excess may interfere with zinc-based enzyme function affecting memory and mental function.
Fluorine	Components of bones, teeth, other tissues.	Occurs naturally in water, but usually from fluoridation	RDA of 1.5 to 4 mg Versus TDR of 2 to 4 mg	Found in bones and teeth and appears to improve tooth structure and inhibit tooth decay.	Deficiency in natural water associated with increased aggression and violent behaviour.

Appendix 2: Continued

Vitamins	Storage & Comment	Major Food Sources	RDA or DV[a] versus Therapeutic Dosage Range[b] (TDR)	Functions	Deficiency/Excess Symptoms and Disorders
Micro-Minerals Found in the Human Body					
Cobalt	Constituent of vitamin B_{12}.	Dietary sources include organ meats and meats.	Unknown	Structural part of vitamin B_{12} and required for erythropoiesis, the formation of new red blood cells.	Seldom deficient as a trace element, but in extreme deficiency may lead to **anaemia**.
Chromium	Found in high concentrations in brewer's yeast.	Dietary sources include brewer's yeast, wines and some brands of beer.	RDA of 50 to 200 µg Versus TDR of 200 to 300 µg	Needed for normal activity of insulin in carbohydrate and lipid metabolism. Key component of glucose tolerance factor increasing effectiveness of insulin and thus lowers blood sugar levels.	Deficiency may result in lower energy levels, and apparent heart conditions that improve with supplementation.
Selenium	Recognised for many years as cause of disease in horses and pigs before realised its importance in human nutrition.	Dietary sources include seafood, meat, chicken, grain cereals, egg yolk, milk, mushrooms, and garlic.	RDA of 50 to 70 µg Versus TDR of 200 to 300 µg	Powerful antioxidant and free radical scavenger protecting liver, brain and muscles as part of enzyme glutathione peroxidase which protects cell membranes from free radical damage.	Deficiency results in white muscle, liver and pancreatic disease in animals especially pigs and horses and **Keshan disease** in humans. Keshan disease named after a region in China, and causes a type

Appendix 2: Continued

Vitamins	Storage & Comment	Major Food Sources	RDA or DV[a] versus Therapeutic Dosage Range[b] (TDR)	Functions	Deficiency/Excess Symptoms and Disorders
					of heart disease in which there is weakening of the heart muscles.
Micro-Minerals Found in the Human Body					
Selenium continued	Some soils very deficient limiting selenium in crops grown in these soils. This is especially true in Australia where it is essential to provide livestock selenium for normal health.			Co-factor in enzyme needed in vitamin E metabolism, another antioxidant. Prevents chromosome breakage and may play a role in preventing certain birth defects and even cancer.	
Trace Elements: Minerals only found in trace concentrations in the human body, yet have been shown to affect certain functions					
Molybdenum		Only in trace amounts from foods.	25 to 75 μg No RDA just amount need for optimal function.	Required for the function of several enzymes.	
Nickel		Only in trace amounts from foods.	Unknown	Aids iron absorption, needed for optimal growth and reproduction in animals.	

Appendix 2: Continued

Vitamins	Storage & Comment	Major Food Sources	RDA or DV[a] versus Therapeutic Dosage Range[b] (TDR)	Functions	Deficiency/Excess Symptoms and Disorders
Trace Elements: Minerals only found in trace concentrations in the human body, yet have been shown to affect certain functions					
Silicon		Only in trace amounts from foods.	Unknown	Required for bone growth and connective tissue development in animals.	
Arsenic		Only in trace amounts from foods.	Unknown	Required for adequate growth and reproduction in animals.	
Boron		Only in trace amounts from foods.	Unknown	Enhances parathormone action and the metabolism of calcium, potassium and magnesium.	

[a] RDAs given as ranges, because women and men often require different amounts and American and European standards differ for the same nutrient with higher concentrations in the United States.
[b] Therapeutic Dosage Range derived from the literature and *Nutritional Influences on Mental Illness: A Source Book of Clinical Research*, by Mervyn Werbach.

Appendix 3

ThinkingAdvantage:
One Solution to Optimizing Mental Performance

ThinkingAdvantage: One Solution to Optimizing Mental Performance

Introduction

Clearly to develop a successful nutriceutical to optimise brain function requires answers to all of the questions in Chapter 9. To do this requires an effective tool to get direct feedback to know which of several possible forms of each component truly optimises mental performance. Direct biofeedback can also determine exactly how much of each component synergises the other components in the formula to create a fully functional synergistic matrix.

Applied Neurology: A Powerful Biofeedback Tool

Applied Neurology is the use of muscle-acupressure feedback to assess nutrient-brain interactions. Although not commonly known, it is one of the best biofeedback tools presently available. Applied Neurology can be used to evaluate if nutritional deficiencies affect specific brain functions such as frontal lobe working memory, the seat of many of our executive functions, and hippocampal short-term memory functions.

Furthermore, if deficiencies are present, the same biofeedback can determine exactly which nutrients, including both the form and amount, are required to re-establish full brain function, and whether this nutrient combination is actually synergistic.

Direct biofeedback can also show which 'key' nutrients are necessary to maintain integrated brain function even when people are under stress, especially the stress of decision-making. This feedback has also delineated the composition of the complex nutrient matrix required for optimum utilization of these key nutrients. Without this information any product is pretty much 'hit or miss'!

If the person happens to have a complete nutrient matrix available in their brain, then the key ingredients present in the product will work and produce an obvious effect – a 'hit'. If on the other hand, the person taking the product is marginally deficient in one or more of the nutrients required in the matrix, even though the product may have all of the 'big guns', there is no way to 'fire' them – a 'miss'!

THINKINGADVANTATGE: AN EFFECTIVE NUTRICEUTICAL TO OPTIMISE MENTAL PERFORMANCE

ThinkingAdvantage is a nutriceutical that was developed using a combination of direct biofeedback, the scientific literature *and* on over 15 years of clinical research using Applied Neurology aimed specifically at enhancing brain function and mental performance. This means seeing what actually worked in practice, not just in theory.

APPENDIX 3

While the scientific literature provides a guideline of what nutrients 'might' work, only clinical experience actually tells you which nutrients actually 'do' work and what works best! And of equal importance, only clinical experience indicates for what percent of the population, and for what percent of the time these nutrients will be effective.

This unique direct biofeedback technology permitted a direct measure of both the nutrient concentration and form to create and maintain integrated brain function, especially under stress. Thus, the optimum concentration, form and synergy of each nutrient in the formula was determined not from theory or guesswork, but directly from biofeedback and years of experience of what actually worked in clinical practice.

Many years of clinical experience with thousands of people demonstrated that when a nutritional product re-established integrated brain function as evidenced by this biofeedback, taking the product actually enhanced mental performance in almost every case. On the other hand, when biofeedback showed no response to a nutritional product, even one that in theory should work brilliantly, taking the product had absolutely no effect on mental performance or re-integration and maintenance of brain function.

Using this technology we also assessed whether loss of integrated brain function resulted from marginal nutritional deficiency or some other factor. Clearly, only the loss of brain integration and mental performance based on nutritional deficiency will respond to nutritional supplementation, as there are a number of other factors that could affect mental performance. Surprisingly, in a large percent of cases, marginal nutritional deficiencies played a major role in loss of integrated brain function and hence mental performance.

Often the only difference between the product that worked and the one that didn't was the lack of adequate concentrations of one or more of the key ingredients, or lack of one or more of the ancillary nutrients in the nutrient matrix, or the lack of one or more key ingredient in an effective form.

Each component of ThinkingAdvantage was therefore tested directly for its ability to re-establish integrated brain function, both singly and in combination with the other components. In this way, we have ensured that each component effectively synergises all of the other components.

ThinkingAdvantage: Safe yet Effective

Likewise, any adverse reactions to the components of the product were checked for during development of the product. In all cases where marginal nutritional deficiencies were found to affect integrated function, ThinkingAdvantage caused rapid restoration of integrated brain function and enhanced mental performance based on direct biofeedback, and personal case histories. In only a very small percent of cases were any adverse reactions noted, and these were generally mild and transitory.

What ThinkingAdvantage can do for You

ThinkingAdvantage will help maintain your brain in the Frontal Lobe 'turned on' and Limbic resistance 'turned off' mode, even when under considerable stress providing optimum problem-solving and decision-making abilities. Of equal importance, your physical body will be under considerably less physiological stress due to reduced activation of the Fight or Flight reactions reducing overall adrenalin and cortisol levels.

The Limbic 'full on' mode and Frontal Lobe 'shut down' mode is characterized by strong survival emotions of the Fight or Flight reaction. These survival emotions are associated scientifically with high levels of physiological stress, resulting in strong activation of the energy consuming sympathetic nervous system, and high levels of the stress hormones cortisol and adrenalin that negatively affect our physiology (e.g. increasing our heart rate, blood pressure, suppressing the immune system, etc.) and interfere with mental performance and memory.

Indeed, chronic high levels of sympathetic activity and stress hormones are strongly associated with a number of chronic illnesses such as gastritis-ulcers, colitis, high blood pressure, stroke and coronary heart disease. So in theory, using ThinkingAdvantage has the potential not only to improve your mental performance, but also reduce the threat of chronic stress-related disease.

ThinkingAdvantage was designed to assist people to maintain frontal lobe functions, especially Working Memory (the home of our executive functions), balance Limbic survival emotions, and enhance memory functions even when people were confronted with their most stressful situations. Therefore, taking ThinkingAdvantage may:

- Allow you to stay calm and prioritise what is important.
- Increase focus and ability to stay on task.
- Increase memory and ability to multi-task.
- Increase ability to access lateral thinking and creative problem-solving.

- Increase ability to learn and implement new problem-solving strategies.
- Reduce stress levels and potentially prevent or reduce stress-related illness.

ThinkingAdvantage and Effective Training

While ThinkingAdvantage offers all of the above significant advantages, one of the key functions of this formula is to keep your brain in optimum learning mode. It is important to emphasize at this point that having full access to the Frontal Lobe functions, although a prerequisite for using your executive functions, does automatically give you effective tools and strategies of how to best use these functions.

This is the role of training programs on lateral thinking, creative problem-solving and effective decision-making skills. Because ThinkingAdvantage helps to maintain frontal lobe resources, balance Limbic emotions and enhances access to memory, it provides the foundation upon which to implement the skills taught in a number of excellent training programs, such as the programs of Franklin Covey, Anthony Robbins and Eduard DeBono that provide effective tools and strategies for optimizing your problem-solving skills and mental performance. This can give you a strategic advantage in both your personal and business environments.

The role of Nutrition plays in Brain Integration that underlies Optimal Mental Performance supporting both Psycho-emotional balance and Effective Training is summarized in the Table 1 opposite.

APPENDIX 3

Table 1 The Three Pillars or Foundations Optimal Mental Performance

Biochemical – Nutritional Foundation	Brain Integration Foundation	Training or Strategies to Use Foundation
The Problems Blocking Optimal Mental Performance		
The foundation of all brain function is nutritional. Only with optimal levels of nutrients that provide precursors for neuro-transmitter, receptor & transporter production and maintenance, and the complete matrix of nutrients required to support this production can the brain function at optimal levels.	The second requirement is complete integration of neural flows. Because the brain is a multiplexed multi-modular structure whose function is totally dependent upon the precise timing and synchronization of neural flows. It is only with full and complete integration of all brain functions that optimal mental function can be achieved.	The first two Pillars only give you the 'potential' to establish optimal mental performance, they do not give you the skills and strategies necessary to do so. This requires training and education in how to implement effective and successful techniques and strategies to truly optimize mental function.
Most people for dietary and/or genetic reasons suffer varying degrees of 'marginal nutritional deficiencies' that 'block' to varying degrees optimal mental function when they are under stress. Only if optimal levels of nutrients are available when you need them, especially under stress, can you achieve optimal mental performance.	Some of the major factors desynchronizing neural flows, and hence resulting in loss of integrated brain function are psycho-emotional factors that trigger our survival emotions and shift the brain from logical or creative thinking into 'fight or flight' reactions. This 'blocks' our solution-oriented thinking. Many of these often have childhood origins.	However, without the first two pillars required for optimal mental function, the third pillar, Training, is often less effective or even ineffective in achieving its potential. Due to the 'transfer problem' – the failure of many participants to implement the fine skills & strategies they were taught to their actual work environment because of loss of Brain Integration.
The Solutions providing Optimal Mental Performance		
ThinkingAdvantage: ThinkingAdvantage is a nutriceutical designed to maintain integrated brain function when a person is under considerable stress. It was developed via our proprietary biofeedback technology such that each component works syner-gistically with every other component.	**Basic Brain Integration and Brain Integration under Stress:** These are two tools designed to provide and maintain Brain Integration (BI) even under considerable psycho-emotional stress developed by our proprietary biofeed-back technology – Applied Neurology.	**Natural Strategic Alliances:** This is the role played by the many fine training programs in the business world that teach these skills and strategies, such as: Anthony Robbins Train-ings and Programs; the DeBono Institute Trainings and Programs; The Covey Institute Trainings and Programs; to mention but a few of many.

253

Table I continued

Biochemical – Nutritional Foundation	Brain Integration Foundation	Training or Strategies to Use Foundation
ThinkingAdvantage is currently the only nutritional formulation containing: all the dietary precursors for the major neurotransmitters, receptors and transporters involved in brain function; all the ancillary nutritional co-factors & co-enzymes and other nutrients required to convert these precursors into 'active' molecules; the correct 'form' of these essential nutrients; and the correct concentrations of these essential nutrients to provide a solid biochemical foundation for optimal brain function!	If the survival emotions have been triggered strongly due to psycho-emotional factors associated with any business situation, the frontal lobe executive functions may be 'shut-down' resulting in the loss of our problem-solving skills and lateral/creative thinking abilities when we most need them. Basic BI & BI under Stress can defuse these psycho-emotional stresses underlying the loss of synchronised mental function when confronted with stressful business situations through effective acupressure and emotional stress defusion techniques.	All of these Training programs, even the highly regarded training programs, suffer from the 'transfer problem' created by not having the first two pillars in place before receiving training. Hence these programs are the natural strategic allies of Thinking Revolution, Inc. whose products can provide robust brain integration that can be maintained even under considerable stress in their work environment, allowing people to now implement the effective creative, problem-solving strategies learned in their training programs.

THINKINGADVANTAGE AND ZINC: THE MISSING LINK

While ThinkingAdvantage can help maintain brain integration even in the most stressful situations, chronic zinc deficiency can compromise the effectiveness of taking this supplement. This is because zinc based enzymes underlie many hippocampal memory and frontal lobe executive functions and therefore, zinc deficiency may disrupt both thinking and memory.

So when people have an overt zinc deficiency, they may experience reduced or even no results from supplementing with ThinkingAdvantage. In the 20% to 50% of people in many Western populations who are zinc deficient, ThinkingAdvantage may be less effective at maintaining their brain integration when under stress, unless the zinc deficiency is also addressed.

How Many People Need Zinc And How Do You Know?

Clinical observation of hundreds of people taking Thinking-Advantage suggests that generally 80% or more of people realise a benefit from supplementing with this effective nutriceutical. However, the discovery that chronic zinc deficiency may compromise the effectiveness of taking ThinkingAdvantage means that it is important for people to be aware of their zinc status.

But how do you know if you are zinc deficient?

Fortunately there is a relatively easy way to assess your zinc status – you simply taste a solution of 0.1% zinc sulphate. A few drops of this zinc test solution are placed on your tongue and you 'taste' it by moving your tongue around inside your mouth. Within 10 to 30 seconds you will either 'taste' the solution or there will be no taste, it tastes like water.

In fact there are four broad categories of taste, with each taste category representing your current zinc status. These tastes and the respective status of zinc in your body are given in Table 8 below.

So clearly, it is important to first test your zinc status before supplementing with ThinkingAdvantage or other nutriceuticals to enhance mental function, as chronic zinc

Table 8 Responses to Tasting a Zinc Taste Test Solution

Taste of Zinc Test Solution	Zinc Status
Tastes – bitter or metallic, and coats your tongue and mouth and taste persist for at least half an hour to over an hour.	Adequate zinc levels in your body.
Tastes – slightly bitter or metallic, or of 'something', but taste disappears within a few minutes.	Borderline Zinc deficiency.
Tastes – nothing, like water.	Chronic Zinc deficiency.
Tastes – sweet or good.	Severe Zinc deficiency.

deficiency may well negate much of the positive benefits of this supplementation.

Sachets of free Zinc test solution are available from the ThinkingAdvantage website at *www.thinkingadvantage.com* should you wish to assess your current zinc status.

ThinkingAdvantage Organic Zinc: A Complete Solution to Chronic Zinc Deficiency

As discussed in Chapter 4, zinc deficiency particularly affects frontal lobe and hippocampal memory functions in the brain and because the brain rapidly takes up zinc gluconate, supplementation with zinc gluconate would seem sufficient to address this problem. However, without a proper nutrient matrix, supplementation with any zinc supplement is often ineffective. Also, chronic zinc deficiency not only affects brain function, but has widespread affects on many systems in the body. It:

- Disrupts frontal lobe executive functions and memory.
- Suppresses the immune system with increased tendency towards infection.
- Reduces effectiveness of liver detoxification system, especially heavy metal detoxification.
- Reduces effectiveness of digestive enzymes resulting in poor protein digestion.
- Reduces wound healing, and often causes cracked skin, brittle nails and dry hair.

ThinkingAdvantage Organic Zinc was also developed using biofeedback technology to address not just the effects of zinc deficiency on the brain, but as a comprehensive solution to zinc deficiency in the body. Therefore, *Organic Zinc* contains various forms of zinc needed to support all affected systems, plus the nutrient matrix to synergise uptake, assimilation and utilisation.

If you discover that you suffer from a chronic zinc deficiency, it may be prudent to take *Organic Zinc* for a month or more until you can taste the zinc test solution to address this condition at least while, you are taking *ThinkingAdvantage* to optimise your mental performance. *ThinkingAdvantage and ThinkingAdvantage Organic Zinc* are available from: www.thinkingadvantage.com.

ThinkingAdvantage the Thinking Person's Solution

In the current world of high stress levels, rapid information processing, and rushed decision-making, *ThinkingAdvantage and ThinkingAdvantage Organic Zinc* can significantly enhance the quality of your decision-making and level of mental performance.

But they can also do much more than that, because just by keeping you out of Fight or Flight survival reactions, your overall stress levels are greatly reduced, which turns off the physiology of stress. This not only leaves you feeling mentally calmer, but also reduces biological wear and tear on your body.

ThinkingAdvantage products provide a one-stop nutriceuticals to fully address the effects of stress on mental function, and assist you to stay in optimum learning and performance mode. They have been proven to provide the ability to maintain brain integration during periods of heavy mental demands, tight deadlines, long working hours and high levels of emotional stress, whereas without these products, this would not have been possible. Read the testimonials below for support of this statement.

Testimonials

Congratulations!! Your ThinkingAdvantage seems to work fantastic!! My brain sucks it in and spits out good work! This is why I'm still taking it today two years after I was first introduced to it!
Alfred Schatz, Director, Institute für Angewandte Kinesiologie, Freiburg, Germany 17-6-03

I have a ThinkingAdvantage success story. I had to take the Series 24 General Securities Principal examination in relation to my role in our socially responsible investment on-line trading company. This is a hard test, which was made harder after the securities analyst scandals. To pass, one needs to get 70% of 150 questions right, all dealing with different very specific securities laws and rules in relation to tricky situations. According to the instructor in the cram class I took, pass rates have dropped to well below 50% nationally. He said people from his classes were passing but with percentages in the low 70s.

I have not taken a serious test since the Series 7 Registered Representative exam, which I took in 1988. I studied the book for this test and took the practice exams, which were pretty good preparation but It turned out that his class was off the mark in relation to the version of the test that I got and little of what I learned in his class helped. I was faced with a lot of difficult questions with answers that weren't obvious. However I doubled my dosage of ThinkingAdvantage and managed to work my way through the exam, coming out with an 85%.

I am clear that passing would have been much more difficult without ThinkingAdvantage, and I know that I would have gotten much more nervous without it. Not only that, but my memory also seemed sharper and I seemed to be able to reason better. This is not a very scientifically verifiable success story, but I'm clear about it and it was important to me. So, thanks again for developing ThinkingAdvantage.
James Hurd-Nixon, CEO of Sustainable Systems and Progressive Trade, San Francisco, California, October 2004.

'Thank you very much for sending us ThinkingAdvantage – it seems to make miracles, we both take it every day and everything is going much easier, we both feel much better.'
Andrea and Charly Hahn, Directors of Lernzentrum, LEAP

Practitioner and High School Teacher, Linz, Austria, November 2004

'At all times my husband came home after work badly tempered and started complaining about anything for about a quarter of an hour (instead of saying he was tired of work and hungry). Furthermore it was difficult to discuss his job situation or anything with him in a rational way. Now since November 2003 he has taken ThinkingAdvantage and he has changed to a normally friendly person.

For us this effect is even more important because since January he definitely is drowned in work (one of his colleagues suffered her second brain tumor and she was, like my husband, the leader of about 15 people). Now my husband and his boss share her work and responsibilities as well their own. Thank goodness for ThinkingAdvantage as he is able to behave nicely towards his family and – at least partly – to think clearly about what to omit in his job and what definitely has to be done.'

Mathematics Teacher, Kinesiologists & LEAP Practitioner, Name withheld on request, Munich, Germany, October 2004

I have been taking Thinkingadvantage and it is really good. For over twelve months (the menopause stuff), I have been overwhelmed if there is too much on and the anxiety levels have been high. Just thinking about things that I wanted to complete like my Diploma of Nutrition and Advanced Diploma would send me into an acopic (new word for not coping) spin.

Since taking the TA – my attitude to all that I am doing has become considerable different. I even took on some extra

work to write and teach a medical terminology course at the same time as doing my Counselling Kinesiology assessments and my Advanced Applied Physiology assessments, all while changing roles at the Hospital where I work. I had all of this to do plus teaching a module at College in the next 4 weeks, and yet I calmly know that I will be OK and I will get there. No stress at all! I will continue to take TA.
Kathy Carmurciano, Nurse in Spinal Care, Applied Physiologists and Teacher, February 2004

We bought a bunch of the ThinkingAdvantage at the end of last year (2003). My kids and my wife Nicole and I have been taking it. The kids remarked that they could tell when they forgot to take it because their math classes were more difficult. Thank you for your innovation.
Mick MacKenzie, Author and Developer of Self-Actualised Learning Technique (S.A.L.T.), October, 2004

'I must say I have had a really powerful experience with the integrative power of "ThinkingAdvantage (that's why I'm still taking it)." For example, I was driving my daughter to school in subzero weather and got into a fender binder (my fault). But despite my stress levels I noticed I could still think very clearly! So it really is helping to keep both hemispheres working together!'
Dr Carl M. Anderson, Ph.D., Research Neurologist & Assistant Professor of Psychiatry. January 16, 2004, began taking TA in April 2003

Some ThinkingAdvantage feedback: one of my clients who is troubled with migraine headaches said that she was

beginning to get one the other day, and went ahead and took her ThinkingAdvantage (since it was morning and that is when she usually takes it) along with four Advil to try and avert the onset of the migraine pain. She said that within a matter of minutes the migraine disappeared. She knows it wasn't the Advil as she has taken it many times and it only slightly dulls the headache. She attributed the disappearance of the migraine to ThinkingAdvantage and is delighted. She wanted me to be sure and pass this on to you.
Kate Rupert, Applied Physiologist, Evaningston, Illinois, November 2004

I have had migraine headaches at a frequency of one to two per month since I was six years old (I am now 50). About six months ago I began to take Thinking Advantage (TA) on a daily basis to help me with my cloudy thinking (which it's done wonders for). One morning I felt the old familiar sensation of one of my migraine headaches coming on. My vision became affected with my usual aura, zig zag pattern, flashing lights and tunnel vision indicating that one was on its way. Despite the sensations, I went ahead and took my TA as I prepared to resign myself to another unproductive day in bed. And then something astonishing happened. Within minutes of taking the TA my vision cleared, and the majority of my migraine pain simply vanished. There was nothing else to attribute that change to. And the most astonishing part is that since that day I have not had one single migraine headache. Thinking Advantage has changed my life.
Leona Schmeidler, Housewife and Mother, Salt Lake City, Utah, March 2005

APPENDIX 3

As I've been taking Thinking Advantage for a while now, and I've noticed a couple of things that might be of interest for you: First of all, I have a lot more energy to work, to study, to concentrate – but my brain is still active at night and have difficulty getting to sleep (you must know that my sleeping patterns are easily disturbed), however this has improved since taking TA Organic Zinc. Secondly, my mother just told me that my voice was clearer than she heard it in a long time, and that I haven't stuttered once during the conversation I had with her over the phone. I have been stuttering more or less depending on emotional stress levels all my life.
Leila Parker, Applied Physiology Instructor and Kinesiologist, Lausanne, Switzerland.

INDEX

Absolute deficiencies 2, 115
absorbed, absorption 34, 57
adaptogen 147
active ingredients 25
active site 35
Adenosine Triphosphate (ATP) 49
ADHD 70
Alleles 35, 63
amine group 43
Amygdala 89
ancillary nutrient matrix 61, 116, 119
ancillary nutrients 25, 114, 116, 157
anorexia 72
Anti-oxidant molecules 144
antioxidants 17, 60, 61
anxiety 136
apoenzyme 48
apoptosis 60
aromatic ring 43
assimilated 34

Assimilation 58
ATP 130
ATPase 130
Attend 155
Attention Deficit Hyperactivity Disorder (ADHD) 7
Ayruvedic medicine 147

Barcopa monniera 147
bioflavonoids 25
Brahmi 147
Brain Integration 83, 100, 114, 122
Brain Link 155
Brain Longevity 54
brainstem survival system 115
Brownell, Kelly 28
bulimia 72
by-products 119

Carbohydrates 47
cholesterol 46
chromium 48

INDEX

co-factors 48
coenzyme 48
Cognitive impairment 61
cold storage 11
conditional-essential fatty acids 127
conventional farming 15
Corpus callosum 95, 100
Cortisol 46
Cortisone 46
Copper 61
cytotoxicity 60

Daily Value (DV) 19
Deductive reasoning 96
Diet 9
disease deficiency guideline 20
dopamine 135
DV 19

Eating disorders 61
Ego survival 95
Eijkman, Chritiaan 3
enzymatic reactions 48
enzymes 35, 49, 118
essential nutrients 42
excitatory NT's 135
executive functions 85, 86, 107, 130,

factory farming 14
fast foods 28
fatty acids 60
Fight or Flight 105
Fight or Flight Reaction 88, 106
Focus Factor 155

free radical 59, 144
free radical scavengers 144
Frontal Lobes 110
functional groups 43, 45, 118

GABA 135
Gamma Amino Butyric Acid (GABA) 44
genes 34, 35
genetic modification 17
Genetics 34
Gestalt functions 96
Gingko biloba 147
Glucose-Tolerance Factor 48
Glufosinate ammonium 18
glutamates 135
glutamic acid 135
Glutamine 44, 135
Glycine 135
glyphosate 18
GM food 18
Goldberg, Elkonhon 86

haem 65
Haemoglobin 65
haemopyrrole 65
herbicides 15, 18
high-density lipo(fatty acid)-proteins (HDLS) 45
high-density lipoproteins (HDLs) 81
High-Fructose Corn Syrup or HFCS 32
hippocampus 147
Hydroxyl group 43
hyper-vigilance 136

Hyperactivity Attention Deficit
   Disorder (ADHD) 62
hypochloric 56

Immune dysfunction 61
Inductive reasoning 96
inhibitory NT's 135
insomnia 72
intuition 96

Junk Food 29

Knee-jerk Reaction 108
Kryptopyrroluria 64, 68
Kryptopyrrole 64, 65, 72

Labile Pool 77, 80
Learning Enhancement
   Acupressure Program or
   LEAP® 71
Learning Factor 155
Lifestyle 26
Limbic System 94
Logic 96
Logic functions 96
Loss of Brain Integration 101
low-density lipo-proteins (LDLs)
   45

Macro-minerals 46
Macro-nutrients 40
Magnesium 61
Manganese 61
marginal deficiencies 2, 116
marginal nutrient deficiencies 5
megadose 23
Memory disorders 62

Mental Performance 109, 155
metallothionein 59
methyl group 43
methylphenidate 70
Micro-minerals 41
Micro-nutrients 40, 47
mineral co-factors 142
Modified Atmosphere Packaging
   11, 17
modulatory NT's 136
multiplexing 84
mutation 35
myelination 60

Neuropathy 61
neurotransmitters 36, 60
non-essential nutrients 42
Nootropic Herbs 134, 146
noradrenalin 135, 136
Nucleus Accumbens 135
Nutriceuticals 154
nutrient levels in produce 12
nutrient matrix 50, 114
nutritional deficiency disease
   20
nutritional synergy 134

obesity epidemic 32
Omega-3 fatty acids 124, 138
Omega-6 fatty acids 124, 138
Optimum Mental Performance
   112
organic grown food 14
Otto, Dr. Gerhard 71
Overwhelm 101
oxidative damage 144

INDEX

PAG 88
parallel processing 84
pepsinogen 56
Periaqueductal Gray Substance 88
Periventricular Gray Matter 88
Periventricular Hypothalamic Gray Matter or PHG 88
pesticides 15, 16
Pfeiffer, Dr. Carl C. 64
Phenylalanine 43, 135
phobias 93
Phytates 57, 62
precursor molecule 35
product 119
PVG 88
Pyridoxine 61

RDA 19, 160
receptor molecule 34
Recommended Dietary Allowance (RDA) 19
Reservoir Pool 80
Reservoir Pools 77
Ritalin 70

saturated fatty acids 123
scurvy 3, 22
Selenium 61
Sense of Mind 155
serotonin 136
shelf life 10
Siberian ginseng 148
Sleep disorders 62, 72
Stress 101, 103, 108
structural proteins 35

substrate 118
Sugar 6
Sunflower® Program 71
Super Size Me 7
Supermarket surveys 27
Supplementation 73, 74, 120
Survival System 88
synergistic action 25
synergistic nutrients 61
synergy 50, 134, 152

Therapeutic Dosage Range (TDR) 160
Thiamine 4
ThinkingAdvantage 155
ThinkingAdvantage Organic Zinc 76, 80
Trace Elements 41
transporter molecules 34, 58
transporters 58
Tryptophan 44, 73, 136
type-2 diabetes 32
Tyrosine 43, 136

unsaturated fatty acids 123
Utilization 59
utilized 34

visual acuity 139
vitamin B_1 4
vitamin B_{12} 142
vitamin B_6 73
vitamin C 61
vitamin co-enzymes 142
Vitamins 41

Working Memory 86, 147

zinc arginate 75
zinc citrate 75
zinc oratate 75
zinc picolinate 75
zinc sulphate 75
Zinc-finger proteins 59